The Tribology of Internal Combustion Engines

Collected papers from the one-day seminar *The Tribology of Internal Combustion Engines,* held at the University of Birmingham, UK on 7 November 1996.

The Tribology of Internal Combustion Engines

Organized by the Tribology Group of the Institution of Mechanical Engineers, with the kind support of the IMechE Automobile Division and Combustion Engines Group and The University of Birmingham, UK

MEP

Mechanical Engineering Publications Limited
London and Bury St Edmunds, UK

First Published 1997

ISBN 978-1-86058-071-0

A CIP catalogue record for this book is available from the British Library.

Related Titles

Title	Author/Editor	ISBN
Heat Treatment Processes for the Reduction of Wear in Vehicle Components	A Schreiner *et al.*	0 85298 733 0
Handbook of Vehicle Design Analysis	J Fenton	0 85298 963 6
Automotive Lubricants Reference Book	A Caines and R Haycock	1 86058 049 1
Lubrication and Lubricant Selection	A R Lansdown	1 86058 029 7
Lubricants in Operation	U J Möller and U Boor	0 85298 830 3

For the full range of titles published by MEP contact:

Sales Department
Mechanical Engineering Publications Limited
Northgate Avenue
Bury St Edmunds
Suffolk
IP32 6BW
UK
Tel: +44 (0) 1284 724384
Fax: +44 (0) 1284 718692

Contents

Vehicle environmental legislation: past, present, and future

J J TOULMIN MIMI
Rover Group Limited, UK

THE DEVELOPMENT OF EMISSION LEGISLATION IN THE USA

As a direct result of severe and increasing air pollution in the Los Angeles basin area, California introduced the world's first vehicle emission legislation in 1963. This first step, which limited the emissions from the engine crankcase, was followed three years later, with the first exhaust emission standards. The US federal emission legislation was introduced two years later. The US Federal Clean Air Act which required a 90% reduction in vehicle exhaust emissions was first applied to new motor vehicles in 1968. However, such a massive reduction was not technically possible at that time and probably could not have been measured anyway. The motor industry demanded, and got, a series of waivers and were permitted to comply with much less severe emission standards whilst test procedure, measurement method, analysis equipment and the emission control technology for vehicles were all developed concurrently. By 1974 a reduction of about 50% from the uncontrolled levels of the mid 1960s had been achieved, but significant further progress seemed unlikely without a major breakthrough. The only solution appeared to be a chemical one - the use of a catalyst to promote the oxidation reaction in the exhaust system - anbutd such devices were not durable. Quickly the reason was found : the lead in the fuel was a catalyst poison. The US legislator moved rapidly to require the widespread availability of unleaded fuel, and, for 1975, vehicles sold in California, which still retains its own separate vehicle emission standards, were fitted with the new catalytic converters. The standards continued to be tightened every other year or so until 1983 when the original Clean Air Act demand for a 90% reduction in vehicle emissions was achieved. By then however, probably because of the growth in vehicle traffic, it was found that the national air quality standards were still far from being met. The Clean Air Act has since been amended, the California Air Resources Board have progressively and severely tightened still further their requirements, specifying ever more aggressive standards to be achieved through to 2003. The 90% reduction goal has become better than a 99% goal with zero-emission vehicles part of the legislator's package of measures to deliver clean air.

Californian Passenger Car Exhaust Emission Standards			
Year of Applicability	Exhaust Emission limits @ 50,000 miles - g/mile		
	Hydrocarbons	Carbon Monoxide	Oxides of Nitrogen
1972	3.2	39	3.2
1980	0.41	9.0	1.0
1986	0.41	7.0	0.7
1993	0.25	3.4	0.4
1996 + (LEV)	0.08	3.4	0.2
1996 + (ULEV)	0.04	1.7	0.2

After 1993, additional standards apply at 100,000 miles!

In addition, the regulations restrict the amount of maintenance that vehicles require, impose mandatory emission warranties, and recall provisions for defective vehicle designs. Conformity of production and in-service testing are also included. Durability of the engine/emission control system must be demonstrated to 100,000 miles. Soon to be introduced are requirements for the vehicle to handle all the fuel vapours displaced during vehicle refueling, which, subsequently, will have to be consumed by the engine which in turn will have an impact on the fuelling calibration of the engine and hence its potential emission output.

THE DEVELOPMENT OF EMISSION LEGISLATION IN EUROPE

In the late 1960s, Europe was not going to left off the band-wagon of vehicle environmental legislation. The task fell to the United Nation's Economic Commission for Europe to develop appropriate legislation acknowledging the differences in topography and climatic conditions to the east of the Atlantic ocean. The requirements introduced by Washington and Los Angeles were considered, for Europe, to be unnecessarily expensive and complex IGNORING THE FACT THAT MOST VEHICLE MANUFACTURERS HAD ALREADY MADE THE LARGE INVESTMENTS TO MEET the US legislation. On four occasions up to 1983, the UN.ECE in Geneva tightened, slightly, the standards, but Europe took more than 15 years to achieve what the USA had done in 7. At that time, it was eventually recognised that the less costly measurement equipment was not satisfactory, and the US measurement method was adopted. Every time the ECE changed the rules for Europe, Brussels took copy-cat action, transposing the Geneva text into an EEC directive. By 1988, Brussels could wait no longer for the slow pace of developments in Geneva, and took the lead in developing European vehicle emission rules. In their hurry, they managed to introduce standards requiring catalysts (on larger cars) before they had ensured that unleaded fuel would be available, but it remained optional for member states to enforce the new standards. In 1989, the first mandatory directive was introduced which was to apply from 1992 but only to cars under 1.4 litres. Then in 1991, a more radical revision to the rules took place. In addition to a new drive cycle over which to measure emissions, many of the long established American practices were adapted to the EEC rules which now, for the first time, included durability and evaporative emission requirements. These rules were approximately equivalent to the 1983 US standards, but the EU, as it had become, almost halved the limits again for 1996 and has recently proposed some far reaching new requirements for introduction during year 2000. Anticipated new standards will be equivalent to about a 97% reduction from 1970 levels and the legislation will focus more on the emission performance of vehicle in service than from prototypes. The evaporative emission test will be much more demanding, and clean fuels will go some way towards reducing emissions from vehicles old and new.

Indicative European Passenger Car Emission Standards (petrol mid sized car)					
Directive Number	Year of first Applicability	Exhaust Emission limits - g/km			
		Carbon Monoxide	Hydrocarbons + Oxides of Nitrogen	Oxides of Nitrogen	Hydrocarbons
70/220/EEC	1971-72	42	no standard	no standard	2.7
74/290/EEC	1975-76	34	no standard	no standard	2.6
77/102/EEC	1977-80	34	no standard	3.6	2.6
78/665/EEC	1979-81	27	no standard	3.1	2
83/351/EEC	1984-86	21	6.0	no standard	no standard
88/76/EEC	1988-91	7.5	2.0	no standard	no standard
91/441/EEC	1992-93	2.72	0.97	no standard	no standard
94/12/EC	1996-97	2.2	0.5	no standard	no standard
Proposed	2000-01	2.3*	no standard	0.15	0.2
Proposed	2005-06	1	no standard	0.08	0.1

* Test procedure changes result in 2.3 g/km being more stringent than 2.2 g/km to the previous procedure.

EMISSION LEGISLATION IN OTHER MARKETS

Japan introduced its own domestic legislation in the late 1970s using US-style test equipment but their own test procedure and drive cycle. Their focus has been on urban emissions of oxides of nitrogen. China and India have also developed their own test procedures, but all other countries that have vehicle emission regulations have adopted past or present US, ECE or EU regulations in whole or in part, or some combination of both! Canada, Sweden and Australia all adopted, in the 1970s, regulations that were based closely on those developed in the USA. Austria, Norway and Switzerland followed suit in the 1980s but have since switched to the requirements of the European Union. Many pacific-rim countries adopted US regulations or Japanese as alternatives. Recently, as a result of discussions initiated by the European Commission, several countries in the Far East now accept European approvals as alternatives to their own domestic legislation.

NOISE

The very first separate directive published by the European Economic Community was the one to limit the noise emissions from motor vehicles. Directive 70/157, published in 1970, covered almost all road vehicles including motorcycles. This directive mirrored the

International Standards Organisation standard number R364 and imposed a range of standards for vehicles. Derogations were permitted for vehicles equipped with direct-injection diesel engines and for those vehicles designed for off-road operation. There were separate standards for commercial vehicles. The ordinary passenger car limit was 82 decibels [dB(A)] whilst that for the heaviest trucks was 91 db(A). The following table illustrates the development of the noise requirements in Europe:-

EEC/EU Directive No.	Date of Introduction	Limit for ordinary passenger cars [dB(A)]	Limit for Heavy Trucks [dB(A)]
70/157/EEC	1972/73	82	91
77/212/EEC	1981/82	80	88
84/424/EEC	1988/89	77	84
92/97/EC	1995/96	74	80
Predicted	2003/2004	71	76

The test procedure which has been slightly refined over the years consists of driving a vehicle at a constant 50 km/hr and then accelerating at wide-open throttle from a point for a distance of 20 m in 2nd gear (and 3rd gear if a 5 speed). The noise is measured at a point 10 m after the commencement of the acceleration at a lateral distance of 7.5 m from the path of the vehicle. At the level of 71 dB(A), the noise generated at the tyre to road interface becomes the dominant factor and noise from the vehicle's induction and exhaust openings are insignificant. However, it remains important for interior noise. Measures to control noise transmission to the interior of the vehicle carry both cost and weight penalties and so reductions of noise at source remain important.

Twenty years ago, the USA spent a considerable amount of time and energy developing unique and very complex test procedures employing a number of different driving modes. However, the US Environmental Protection Agency made an amazing discovery! An evaluation of the magnitude of noise sources indicated that passenger cars were well down the list, quieter than aircraft, railway trains, heavy trucks and many factories. No US federal noise standards have been promulgated for cars.

Other countries, including Australia, Japan and Switzerland have introduced noise regulations. Vehicles meeting EEC or ECE limits usually comply with ease with these other requirements.

ON-BOARD DIAGNOSTICS

Measurements conducted in the USA of the emission performance of vehicles in use showed that the levels of emission control demonstrated by prototype certification vehicles are not always maintained by vehicles in customers' hands. One of the reasons for this discrepancy between actual and predicted emission performance results from individual component and/or system failures caused by a number of possible reasons. Whether these failures are caused by defective initial design, manufacturing errors, a lack of proper vehicle maintenance, misfueling or vehicle misuse or abuse may be considered irrelevant in terms of environmental impact. Whilst warranty returns may show a design or manufacturing defect in newer vehicles of a given model, and whilst annual roadworthiness inspections will eventually reveal individual vehicles which have extremely high emissions, On-Board diagnostics (or OBD for short) offers a much more comprehensive and faster method of identifying faults, howsoever caused, on vehicles usually within a short distance of the fault developing. This obviously dramatically reduces the average detection time and hence will reduce the total amount of high emissions resulting from component or system failures on the assumption that, once detected, the fault will be speedily rectified.

With the use of ever increasingly sophisticated electronic engine management systems, component or system failures can be increasingly difficult for the driver to detect. The failure of some parts of the emission control system or related engine systems will be likely to cause the engine management system to revert to a back-up strategy in which a missing input from a sensor is substituted with an assumed value. These strategies may be so good that, without the aid of an on-board diagnostic system, the driver is unable to detect any change in vehicle operating characteristics and yet the exhaust emissions from the vehicle may have increased significantly.

Initially introduced into Californian emission regulations at the end of the last decade, more advanced on-board diagnostic systems are now required by both US Federal and Californian regulations, and the European Union is set to follow suit with their own regulations to be introduced immediately prior to the commencement of the new millennium.

The systems are designed to alert the driver when the emissions from the vehicle exceed certain threshold values. However, since the only exhaust gas that we can currently measure during normal vehicle operation is oxygen, it is necessary for the system to deduce from the oxygen content of the exhaust before and after the catalytic converter, that emissions are likely to exceed the permitted levels. To achieve this, calibration of the diagnostic system is critical and must take into account the normal production tolerances inherent in all engines. Thus uniformity and consistency of the engine and the emission control system are essential for the accurate operation of the OBD system.

HARMONISATION

Currently there is a massive waste of manufacturer's scarce resource in the duplication of development, testing and certification approvals necessitated by the differing requirements of major markets. In an apparent, and welcomed change in philosophy from the past, the USA and the EU have opened discussions on the harmonisation of vehicle standards and regulations. The European and US motor industries have, for the first time, presented to governments their common proposals for the harmonisation of testing and certification procedures.

These proposals, which were developed by the American Automotive Manufacturers Association (AAMA) and the European Automobile Constructors Association (ACEA), provided the basis for the conclusions to the Transatlantic Automotive Industry Conference on International Regulatory Harmonisation which took place in Washington USA in April this year.

The conference, whose participants included representatives from the EU Council of Ministers, the EU Commission, the European Parliament, the US house of Representatives and the Director of the ECE Transport Division (WP29), was a positive move towards regulatory convergence. It was agreed that a move to harmonisation could lead to significant savings for industry and the consumer by reducing vehicle development costs by perhaps as much as 10%. This would speed the process of development and enable development and capital costs to be spread over a larger vehicle population. The goal is "tested once, accepted everywhere" - provided that the standards are not compromised in the process. Three main categories of initiatives were identified for further development : process, safety and the environment.

Process Issues : The conference recommended the development of:-

- a process for agreeing on functional equivalence of regulatory standards.
- a process for the mutual recognition of regulatory standards and certification procedures.
- a plan for co-ordinating research by both industry and government
- the role of the UN Economic Commission for Europe WP29 as the forum for global regulatory harmonisation.

Safety Issues : The conference developed a comprehensive list of regulatory requirements for motor vehicle safety in the US and Europe and an evaluation of the functional equivalence of overlapping requirements will be undertaken this year. To begin the shift towards harmonisation across the Atlantic, the vehicle manufacturers recommended:-

- co-operative programmes in the areas of common regulatory activity & development
- the development of a technical basis for the mutual recognition of common regulatory requirements including, in the short term; windscreen wiper systems, head restraints, steering column impact protection and seats.
- the development of a technical basis for the mutual recognition of functional equivalence for unique requirements that provide equivalent levels of vehicle safety.

Environmental Issues : The conference recommended two phases of work :
The first to include that the EU and US governments prepare plans to harmonise noise, electromagnetic compatibility and smoke test procedures and to move towards reciprocal recognition of the functional equivalence of emission regulations, the streamlining of the certification process and the harmonisation of fuel and heavy-duty requirements.

The US and European motor industries agreed on ten first principles of global harmonisation:-
1) Commit to global regulatory harmonisation by becoming contracting parties to the 1958 Agreement and participate in the development of new UN-ECE regulations.
2) Work through and strengthen WP-29 to make it a widely recognised body for the development of global vehicle regulations
3) Establish a work programme to contribute to harmonisation where differences exist
4) Expand the harmonisation discussions to cover all countries
5) Establish mutually recognised certification processes
6) Establish the means to incorporate the functional equivalence of alternative regulatory requirements
7) Co-ordinate pre-regulatory research on the development of new requirements to minimise future divergence

8) Avoid developing unique new national (or regional) regulations without adequate justification
9) Improve the public information process for publicising the development of harmonised regulations
10) Encourage a policy of accepting vehicles which fully meet EU ECE or US regulations.

Clearly harmonisation should bring the benefits of designing to a single set of requirements thereby reducing some of the compromises which are inevitable in the design and development process. It may also move the emphasis of the certification or type approval authorities from their interest in new products to those in actual use!

RECYCLING

Growing economic, social and environmental pressures, and possible future legislation from Brussels, are increasing the interest in design for recycling. The impetus comes from the increasing cost of raw materials and the cost of energy consumed during their refinement and conversion. The increasing cost and taxation of landfill provides the end-of-life disposal industry with the incentive to increase the proportion of the vehicle that can be recycled. The traditional method of disposal of end-of-life vehicles (vehicles that have been abandoned, stolen, failed their annual roadworthiness test or been damaged beyond economic repair) has been for the scrap yard to remove the resaleable parts and then sell the residue to a shredder who would extract the remaining metals for recycling, and pass the remaining materials (glass, trim materials, rubbers etc.) to landfill. This process recovers - for further use - about 75% of the vehicle by weight. In the UK alone, nearly 1.5 million cars are scrapped annually. The EU Commission have identified waste streams as a priority for attention and a working group has been set up to draft a future directive. Their proposal is to require 85% of the vehicle to be recycled by 2002 and 95% by 2015. In this context, incineration to produce heat and power is included in the % recycled as energy recovery, but this will be limited to 5%, so 90% of the material must actually be recycled.

A number of member states - including Austria, France, Germany and Holland, have developed national voluntary agreements on recycling in an attempt to avoid the imposition of an EU-wide directive. In the UK, the Automotive Consortium on Recycling and Disposal (ACORD) has been formed by vehicle manufacturers, the scrap and dismantling industries and various government departments. The UK vehicle manufacturers are set to shortly sign an agreement with the UK government to reduce the need for recycling by longer life designs, and to permit a higher recovery rate for materials by careful component design and assembly techniques. Such design features include ease of draining of all vehicle fluids, ease of removal of air bags, air conditioning systems and the battery. It includes; design for disassembly, the avoidance of mixing materials (such as different types of plastic within a single component) which cannot then be separated easily, the labelling of plastics, and minimising material treatments which make them more costly to recycle.

Recent developments in piston assemblies

M WILLCOCK
ATC - Piston, AE Goetze, UK

INTRODUCTION

The major driving force in engine development is emission legislation and Piston assemblies have to operate in harsher conditions to achieve durability requirements a strategy of alternative design solutions has been developed. The emissions currently monitored are;

i. Hydrocarbons (HC),
ii. Carbon Monoxide (CO),
iii. Oxides of Nitrogen (NOx), and
iv. Particulates (diesel only).

Fuel economy is a selling factor for vehicle manufacturers and a benefit of this is reduced Carbon Dioxide (CO_2) emissions. This is an inevitable by-product of combustion, but improved fuel utilisation and better efficiency can lead to improvements. For the piston assembly, friction is the main factor. It is claimed that up to 44% or more of the total frictional energy can be assigned to the piston assembly alone.

GASOLINE PISTON ASSEMBLIES

To reduce emissions Engine manufacturers are programming the engine management to ensure maximum catalyst efficiency. This can cause transient operation in detonation and overfuelling. For the piston assembly a reduced top land height is the favoured design strategy adopted to reduce crevice volumes and therefore hydrocarbon emissions. This also tends to lead to lower lubricating oil consumption resulting in further HC emissions reduction.

PISTON ASSEMBLY DEVELOPMENTS

GASOLINE PISTONS

- RING / GROOVE INTERACTIONS
- MATERIALS DEVELOPMENT
- DESIGN STRATEGY FOR LOW EMISSIONS

DIESEL PISTONS

- INCREASED POWER OUTPUT
- DURABILITY ISSUES

GENERAL ISSUES

- FUEL CONSUMPTION

Fig 1

HYDROCARBON EMISSIONS

ENGINE NEEDS

- ABNORMAL ENGINE MANAGEME (TRANSIENT HIGH ADVANCE)
- REDUCED CREVICE VOLUMES
- HIGH TOP RINGS (HIGHER TEMPERATURES)
- SMOOTHER BORES (REDUCED OIL FILM THICKNESS)
- OPTIMISED REVERSE GAS FLOW THROUGH RING PACK

POTENTIAL PROBLEMS

- SIDE-FACE WEAR (RING WELDING)
- RING PERIPHERAL SCUFFING

Fig 2

The measures adopted above can lead to a harsh operating environment for the piston assembly. Wear and damage at the piston ring and groove interface, and at the ring to bore interface can occur in these conditions.

RING / GROOVE INTERACTION

A problem often referred to as ring weld describes the damage on the top ring groove due to interaction with ring side faces. This can be attributed to:

(i) reduced levels of lubrication, and

(ii) increased temperature of top ring and groove

These conditions can occur because of the actions taken to reduce emissions, reduction in top land height and engine management program.

The problem is application specific and only occurs in a few engines. However, results to date show that ring welding may be an early life problem and does not always lead to failure. There can be damage during early life, which ceases at a point, so that it may go unnoticed for the remainder of the life of the engine, Figure 3. If damage continues, Figure 4 demonstrates excessive ring pounding against the grooves in a tested piston. Early signs of this problem are high oil consumption and blow-by.

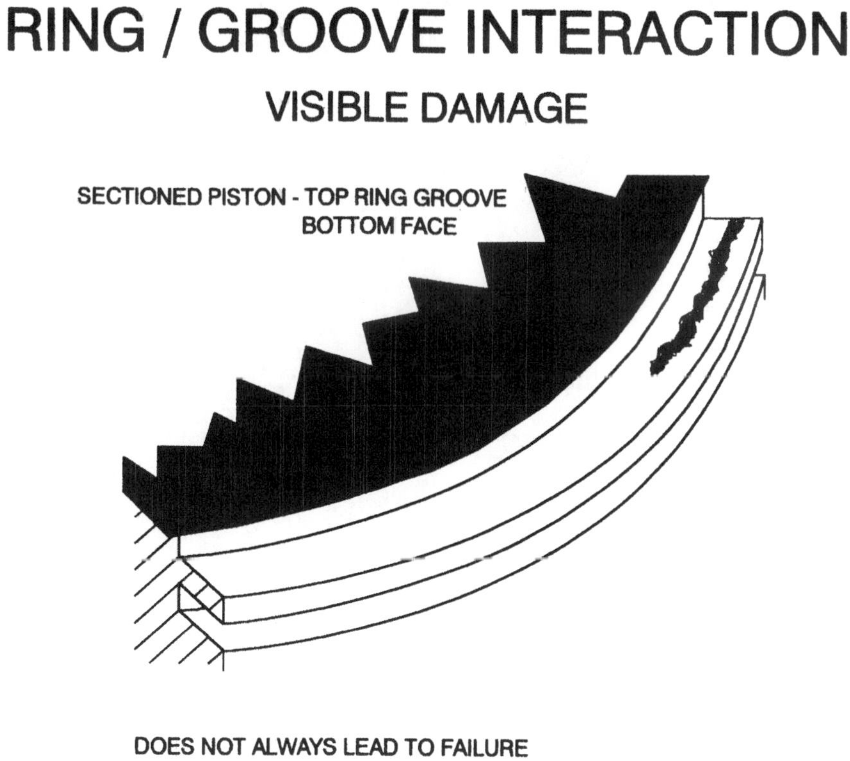

Fig 3

RING / GROOVE INTERACTION

SIDE VIEW OF POUNDED OUT RING GROOVE

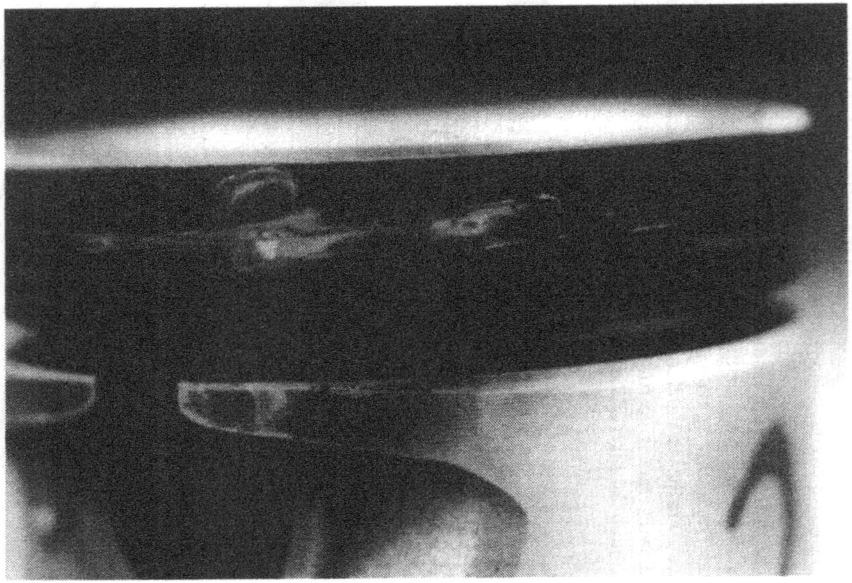

Fig 4

Scanning electron microscope images can show the localised damage on the groove surface, Figure 5. Taking sections through the piston, Figure 6, gives a closer view of the damage to the material. The mechanism for this type of damage is probably a combination of localised fatigue and delamination wear. This occurs when the lubrication fails and the ring comes into contact with asperities on the groove surface leading to localised high contact pressures.

RING / GROOVE INTERACTION

DAMAGE TO TOP RING GROOVE BOTTOM FACE

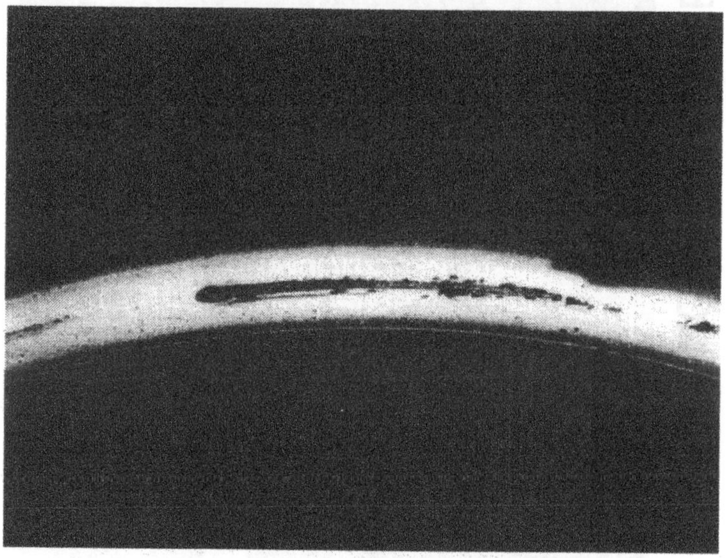

OPTICAL MICROSCOPE IMAGE

Fig 5

RING / GROOVE INTERACTION

EVIDENCE OF DAMAGE

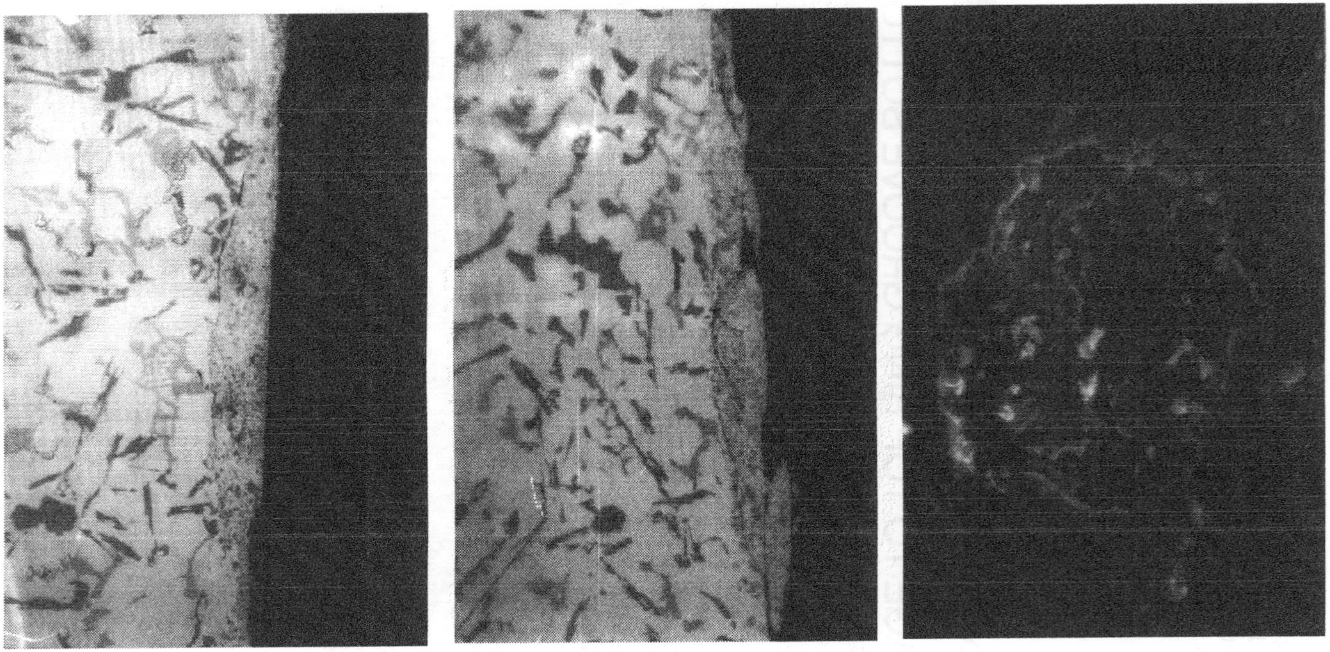

X 500 X 500 X 100

MATERIAL ABOVE CRACK, HAS BEEN EXTENSIVELY DEFORMED
PLAN VIEW SHOWS LAYERS OF ALLOY HAVE FLAKED AWAY

Fig 6

RING / GROOVE INTERACTION

PISTON MATERIALS AND COATINGS AGAINST 13 % CHROMIUM STEEL RING

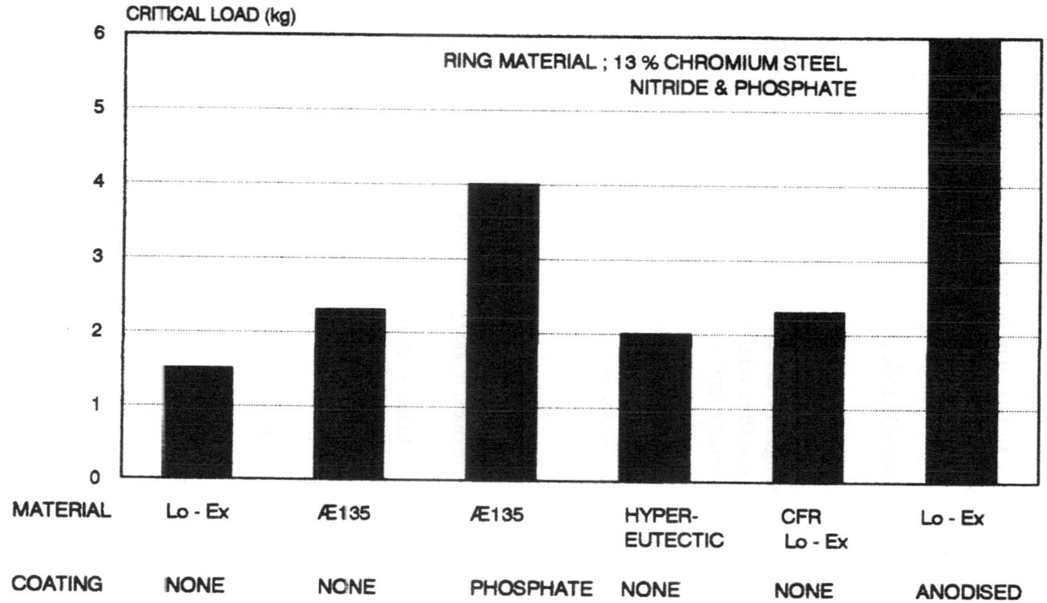

Fig 7

MATERIALS FOR GASOLINE PISTONS

In order to develop materials for resistance to ring weld, a series of tests were conducted to compare different piston and ring surfaces. The apparatus used for achieving wear was a reciprocating wear machine. A pin made up of the piston material including any coatings, was reciprocated against a sample of the ring material. The load and temperature for the test could be varied and marginal lubrication was supplied. The results of these tests are shown in Figure 7, for a range of piston materials. The high copper alloy, AE135, and ceramic fibre reinforced materials showed better wear characteristics running against a standard nitrided high chromium steel surface. The benefit of a phosphate coated surface is also clearly evident. An anodised surface has performed exceptionally well in rig tests and has been successfully employed in the USA for several years now, to counteract potentially serious ring welding problems.

STRATEGY FOR GASOLINE PISTONS

The strategy developed for low emission gasoline pistons is shown in Figure 8. When the top land width is reduced below about 3.5mm, and particularly when detonation is regularly experienced, the strength of the top land becomes the controlling factor. In these extreme cases, copper or ceramic fibre reinforced alloys, which offer high temperature fatigue strength, become necessary to ensure land survival.

Alternative piston assembly designs, which lower the crevice volume with minimised risk of land failure, include a half-keystone ring. These designs are in later stages of development and can enable the top land width measured at the root of the groove to be increased by around 20%.

Recent research is showing that top lands lower than 3.0mm do not achieve significant reductions in emissions, because below this the top land crevice is less significant in comparison to other crevices, ie gasket and valve pockets.

However, if a customer requires a top land width is reduced below about 2.5mm, then additional reinforcement will almost certainly be required. Localised alloying (with nickel for instance) is feasible, but ceramic reinforcement is technically and commercially the more attractive solution for high volume production.

DIESEL PISTON ASSEMBLIES

With each new generation of engine, the power ratings of diesel engines are being increased. The drive for this is the need to reduce emissions. The trends and some of the consequences can be seen in Figure 9. The effect of increased engine power on the piston can be assessed by the thermal loading to which it is exposed, and the trend for increased thermal load is illustrated in Figure 10. In addition to the increased load, direct fuel injection (DI) technology and turbocharging (TC) are being adopted in light automotive applications requiring deep re-entrant combustion bowls in the piston crown. When the thermal load is greater than about 3.8 MW/m^2, the temperatures obtained at the bowl edge can exceed 320°C. Around this temperature there is a significant reduction in the fatigue strength of conventional piston alloys, and this can have a significant impact on the design of the piston. Figure 11 gives the factors to be considered in the design of a light vehicle diesel (LVD) piston.

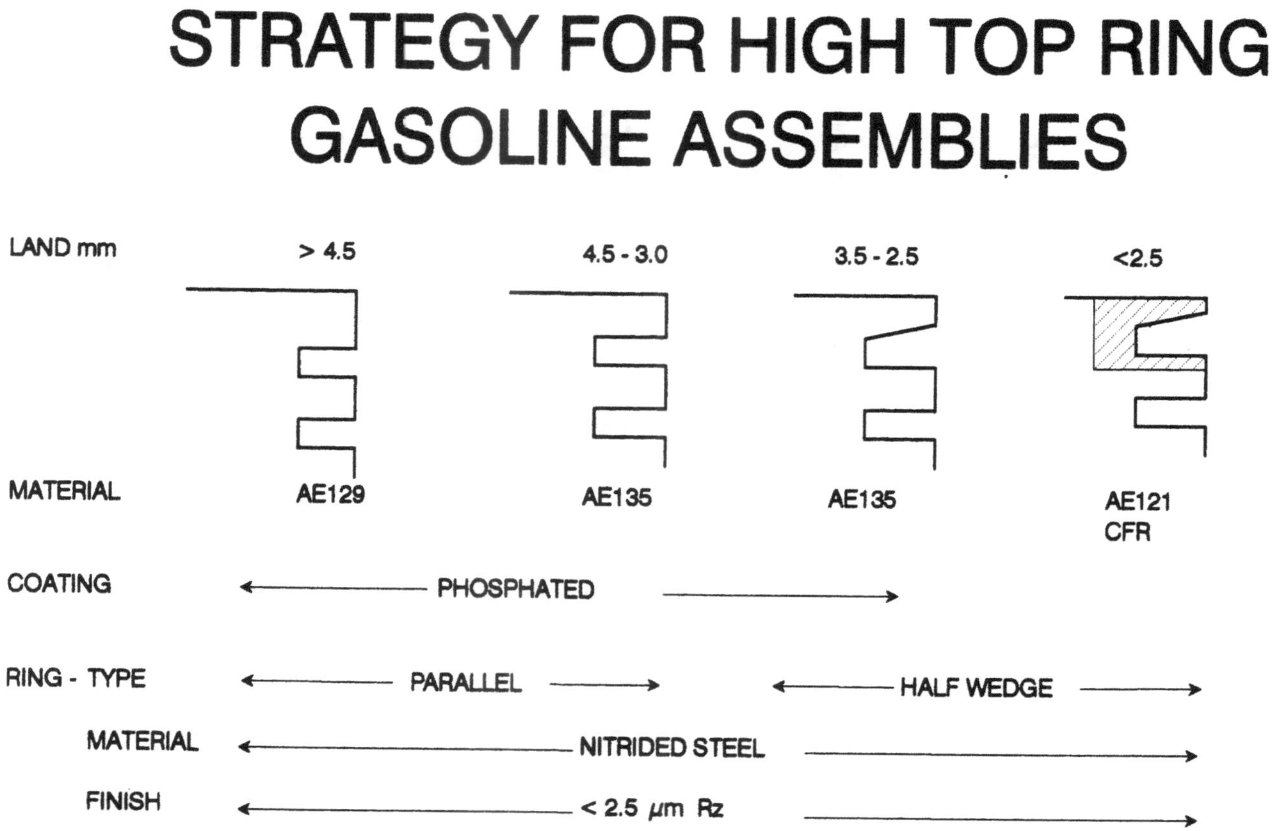
STRATEGY FOR HIGH TOP RING
GASOLINE ASSEMBLIES
LAND mm
> 4.5
4.5 - 3.0
3.5 - 2.5
<2.5
MATERIAL
AE129
AE135
AE135
AE121
CFR
COATING
PHOSPHATED
RING - TYPE
PARALLEL
HALF WEDGE
MATERIAL
NITRIDED STEEL
FINISH
< 2.5 μm Rz

Fig 8

LVD ENGINE TRENDS

EFFECT ON PISTON DESIGN

DIRECT INJECTION	INCREASE COMPRESSION HEIGHT
INCREASING CYLINDER PRESSURES AND THERMAL LOADING	INCREASING BOSS LOAD
HIGHER PRESSURE FUEL INJECTION	BOWL FORM ?
LOW EMISSIONS	LOW OIL CONSUMPTION AND HIGH TOP RINGS
MAINTAIN FUEL ECONOMY DIFFERENTIAL	REDUCED FRICTION

Fig 9

LVD PISTON DESIGN TREND

THERMAL LOAD V POWER

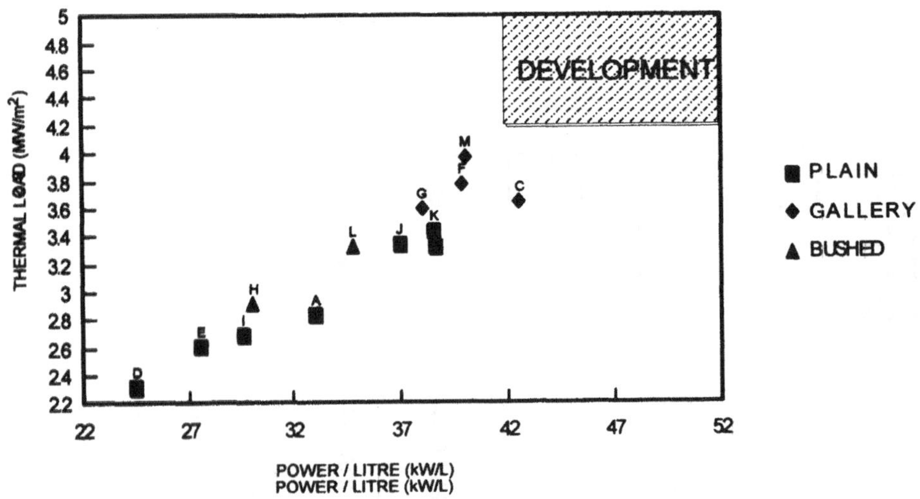

Fig 10

MAJOR FACTORS IN PISTON / PIN

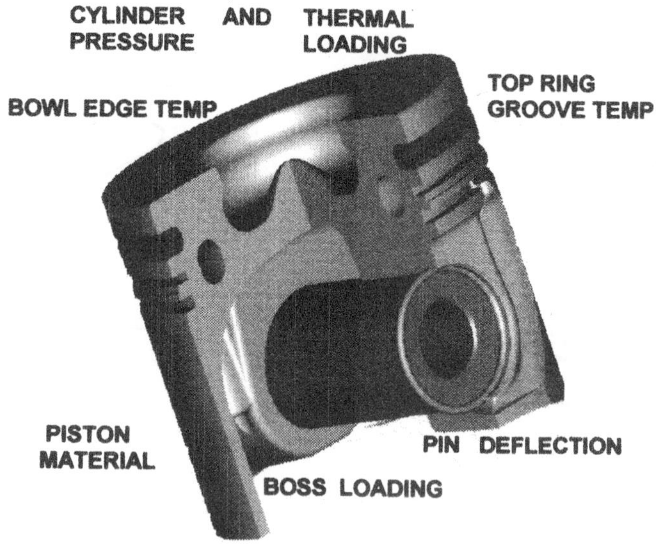

Fig 11

PISTON COOLING

Substantial improvements in crown durability can be achieved by cooling the crown and ring groove region of the piston. To achieve this, oil cooling galleries have been developed for the higher rated LVD piston. By using this method, the temperature can be reduced by 20 or 30°C in this region of the piston. The improvement in the resistance to bowl edge cracking has been confirmed by both suitable engine testing and by analysis based on FE simulation of cyclic loading during the tests. Figure 12 plots the predicted effects of an oil cooling gallery in an automotive diesel piston and shows that the shortest life is at the bowl edge, where an improvement by a factor of 5 can be achieved with gallery cooling.

LVD CROWN DURABILITY

CAST IN COOLING GALLERY

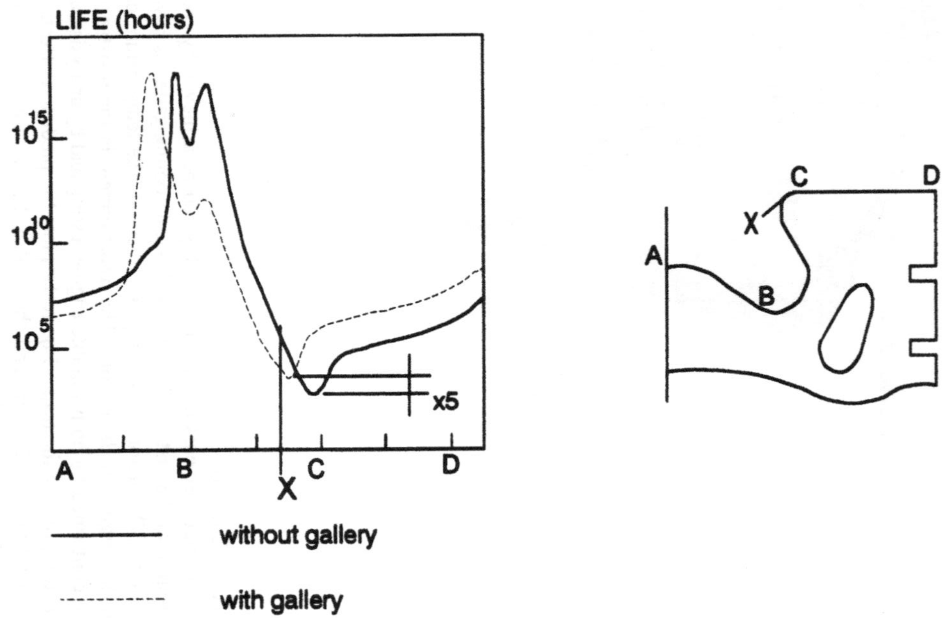

Fig 12

MATERIAL DEVELOPMENT

With the continuing increase in engine ratings materials with improved properties at high temperatures are required to give the necessary durability at raised temperatures, Figure 13 shows characteristics for currently available alloys.

PISTON MATERIAL PROPERTIES

CURRENT ALLOYS

CHARACTERISTIC	Lo Ex	PREMIUM Lo Ex	CFR Lo Ex
THERMAL CONDUCTIVITY (W/ mK)	150	125	110
THERMAL EXPANSION ($\times 10^{-6}$ /K)	24.5	21.5	17
DENSITY	2.71	2.73	2.73
UTS (MPa) @ 350°C	50	55	90
FATIGUE STRENGTH, (MPa) 10^7 CYCLES @ 350°C	32	36	45

Fig 13

Pistons with ceramic fibre reinforced crowns are in series production for a number of highly rated diesel applications, with undoubted durability benefits. Figure 14 shows how Æxelsius 113 can provide a more cost-effective solution. In one turbo-charged application, the substitution of Æxelsius 113 for conventional Lo-Ex has provided an improvement in bowl-edge life by a factor of 3 to 4. As the trend for peak cylinder pressures continues to increase, the mechanical and thermal loadings associated with these pressures will push conventional aluminium materials even closer to their limits, so that piston design will have to be modified in conjunction with alternative high strength materials, Figure 15. Single piece steel pistons are currently being developed for heavy duty diesel applications.

BOWL EDGE DURABILITY

EFFECT OF PISTON MATERIAL

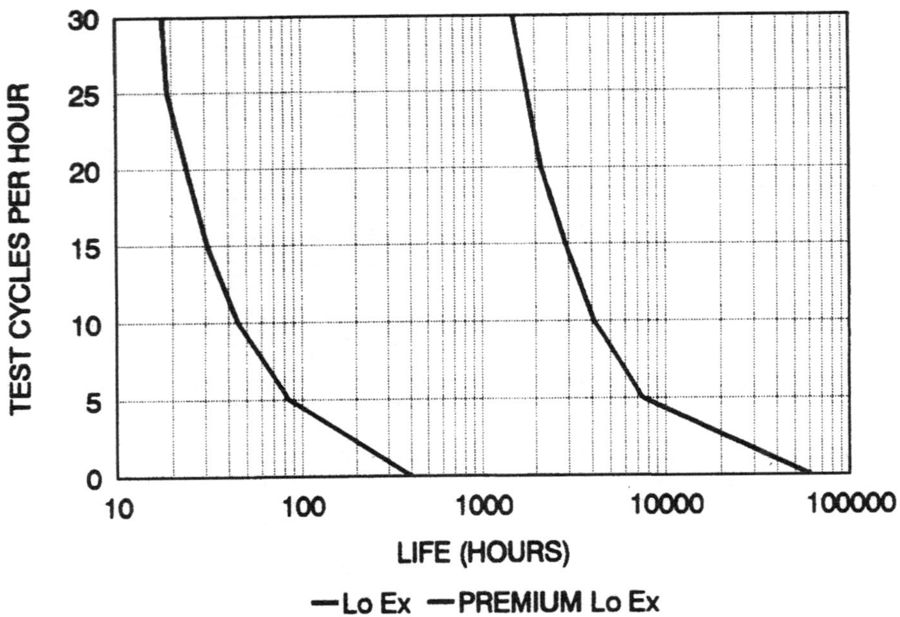

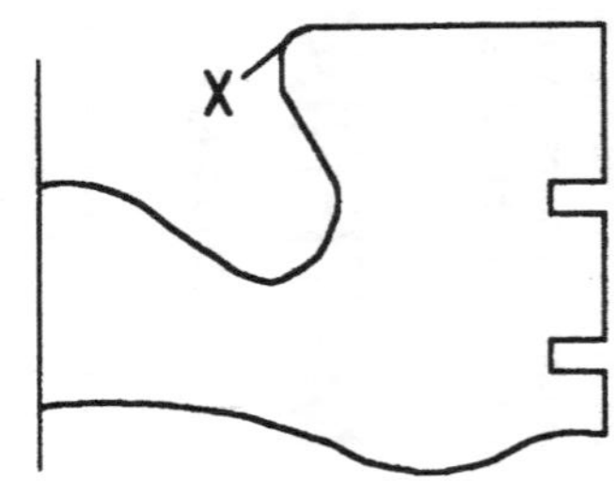

Fig 14

PISTON MATERIAL PROPERTIES

UNDER DEVELOPMENT

CHARACTERISTIC	PREMIUM Lo Ex	PM ALLOY *	S / STEEL**
THERMAL CONDUCTIVITY (W/mK)	125	110	20
THERMAL EXPANSION (x10 /K)	21.5	20	12
DENSITY	2.73	2.8	7.8
UTS (MPa) @ 350°C	55	170	1100
FATIGUE STRENGTH, (MPa) 10 CYCLES @ 350°C	36	70	400

* Al Si 20%, Fe 5%, Ni 2%
** 17 - 4 PH

Fig 15

SPECIFIC FUEL CONSUMPTION

A major factor in the engine friction is the piston assembly, which accounts for approximately 44% of the total. The greatest contributor to this is the ring pack, although the influence of the piston increases with engine speed. Figure 16 illustrates the proportions of these mechanical losses. Several aspects of the piston assembly design have been found to influence friction, Figure 17.

ENGINE FRICTION

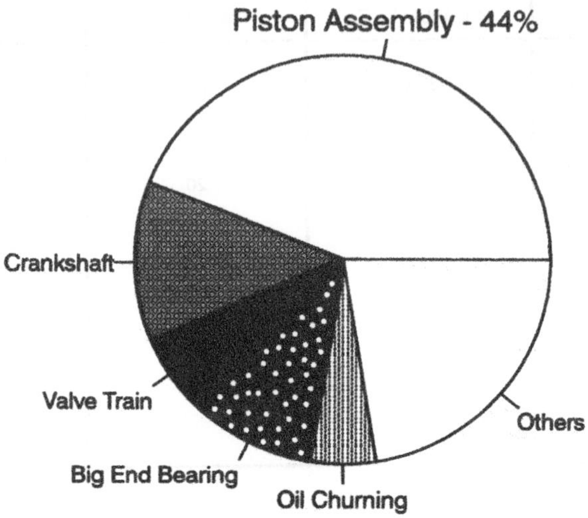

Fig 16

SPECIFIC FUEL CONSUMPTION

REDUCING PISTON ASSEMBLY FRICTION

- REDUCED RECIPROCATING MASS
- REDUCED SKIRT CONTACT AREA
- PISTON SKIRT FINSH / COATINGS
- REDUCED RING PACK FRICTION
    - COMPRESSION RINGS WITH REDUCED AXIAL WIDTH
    - OIL CONTROL RINGS WITH REDUCED TANGENTIAL LOAD

Fig 17

RECIPROCATING MASS

Energy lost due to piston side loads can be reduced by a reduction in piston mass. An additional benefit of low mass is a reduced load on the crankshaft, due to inertia of the piston, and this can allow the down-sizing of the con rod and crankshaft. A shorter compression height (the dimension between the crown and pin centre), can also be achieved to reduce the total height of the engine. However, all piston design factors are inter-related in achieving variations in engine refinement with respect to noise, vibration and harshness. Due consideration must be given to skirt guidance and inter-ring volumes in order to achieve satisfactory noise levels and ring pack performance.

SKIRT CONTACT AREA

Although piston skirt operation is mainly hydrodynamic, boundary lubrication can occur at engine start up, piston reversal around top and bottom dead centre and sudden engine transients. At these conditions, reductions in friction can be achieved by a reduction in the area of piston skirt in contact with the cylinder bore. Two methods of achieving this were developed:

i. the Æconoguide skirt form, which consists of three raised regions on the skirt
ii. the X-piston.

Although reductions in friction have been achieved with both these designs, Figure 18, the Æconoguide form is now most beneficial for diesel applications.

SKIRT SURFACE FINISH AND COATINGS

The specific fuel consumption can be reduced by adopting a truncated surface finish on the skirt. An additional benefit is the stabilisation of piston to bore operating clearance. A low friction coating can have a significant effect on reducing the specific fuel consumption (sfc). These coatings have demonstrated improvements in both sfc and durability, surviving extreme gasoline engine tests. Figure 19 shows the fuel economy benefits of a coating during an ECE-based test cycle.

REDUCED SKIRT CONTACT {AEconoguide}

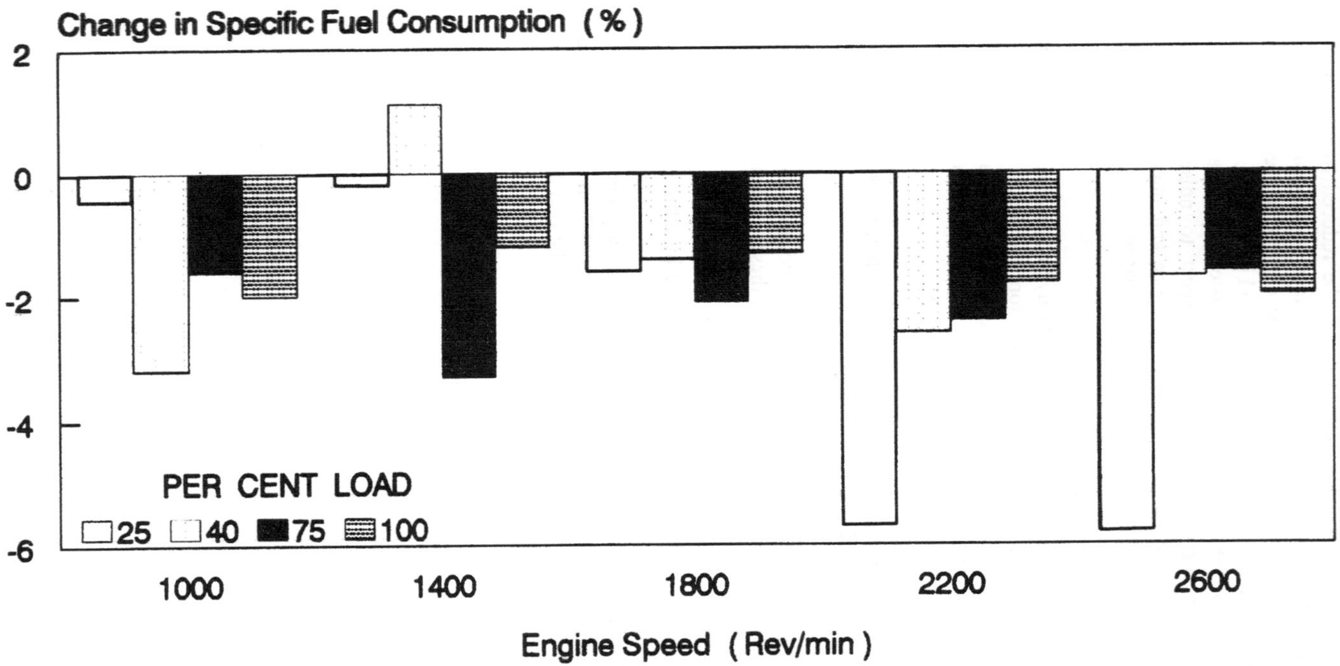

Fig 18

SPECIFIC FUEL CONSUMPTION

SKIRT SURFACE FINISH / COATING

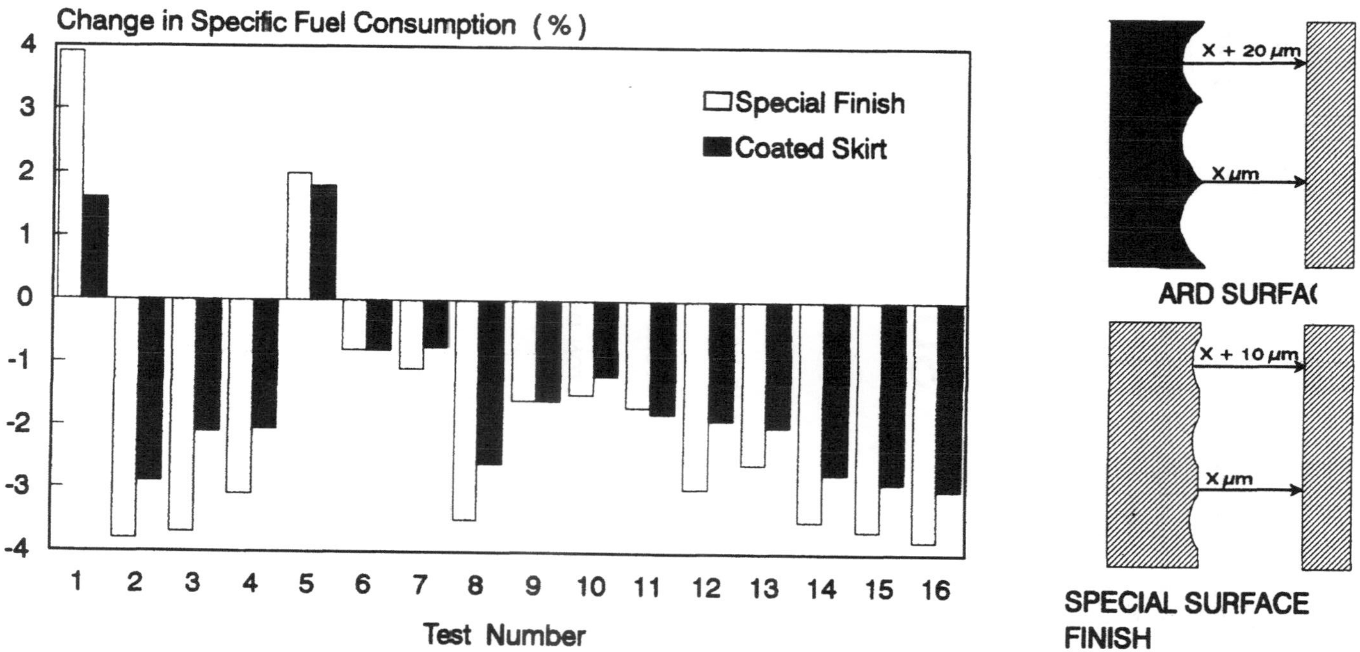

Fig 19

SUMMARY

The continued pressure for low emissions engines has posed some serious challenges to the piston manufacturer. There is some commonality between the gasoline and diesel pistons in the requirement for improved emissions, low oil consumption and reduced friction. These objectives are being achieved with cost effective and durable components.

Low mass and high top ring gasoline piston have been developed and validated for series production capable of achieving significantly reduced emissions. Low friction and reduced oil consumption have also been achieved by improvements in the ring pack.

Increased power ratings in diesel engines have accelerated the development of improved design and reinforcement features to achieve the requisite crown and ring groove durability.

SUMMARY

- IMPROVED ALLOYS FOR BOTH GASOLINE AND CAR DIESEL PISTONS
- STRATEGY FOR GASOLINE PISTONS WITH LOW CREVICE VOLUME
- DURABLE LOW - EMISSION PISTON DESIGNS FOR CAR DIESEL APPLICATIONS
- REDUCTION IN SPECIFIC FUEL CONSUMPTION - ACHIEVED THROUGH REDUCTION IN PISTON ASSEMBLY FRICTION

Fig 20

Cylinder bore surface finish – its evolution and specification

J S LENTHALL
GKN Sheepbridge Stokes Limited, UK

SYNOPSIS

Refinement of the cylinder bore surface finish has been driven by changing requirements placed upon engines over recent decades. Demands to reduce oil consumption with increased durability and resistance to adhesive wear provided the main impetus for change. Evolution of the bore surface finish has been permitted by improvements in honing technology and surface finish measurement. Reasons for refinement, specification of the surface and its manufacture are discussed in this paper.

1. INTRODUCTION

Design trends in the manufacture of internal combustion engines have changed considerably with time. Evolution of the cylinder bore surface finish has advanced more in the medium and heavy duty diesel engine sector than in any other type of engine. Within this sector fuel consumption and durability were the main concerns through the 1970's. In the 1980's these issues were joined by specific output and noise. Engine development in the 1990's has clearly focused on addressing emissions legislation. Bore surface finish and its effect upon oil consumption has affected all these issues. Unburned oil makes a significant contribution to the particulate materials (PM) emitted by diesel engines. Progressing from Euro I through to Euro III diesel engine emissions legislation it is predicted that the PM fraction attributed to oil consumption will reduce from 0.020g kwh^{-1} to 0.015g kwh^{-1}. Most engine builders are experimenting with optimised honed bore surfaces to give this reduction.

Greater interest in cylinder bore surface finish and the first plateau bore specifications emerged during the early 1970's. Before that time cylinder bores were mainly honed using machines equipped with a single expansion head, fitted with one grit size of honing stones. Compared with sophisticated, multi stage honing machines used today, single stage equipment permits only limited variation of the type of surface finish. Before the integrated circuit and microprocessor revolution of the late 1970's, surface finish measurement equipment could determine only a few parameters, mainly, arithmetic mean roughness (Ra), total roughness (Rt) and bearing area slice levels and high spot counts at only a single cutting level. Construction of complete Abbott-Firestone bearing area and amplitude distribution curves was a laborious task in which individual slice levels had to be graphed manually.

From the mid 1970's the wider availabilities of multi stage honing equipment and surface finish measurement machines that could rapidly measure a wider range of parameters, permitted experimentation, measurement and manufacture of more advanced plateau surfaces. Development and specification of the plateau bore surface have been conducted empirically. Throughout this time the manufacture and measurement of the load bearing areas and the distribution and size of oil retaining features at the bore surface have been the goal. Mathematical techniques have been investigated that consider the effects of asperity contact at wear interfaces. However, models that incorporate lubricant retaining features and can optimise their type, size and distribution have not been developed to date. Today many engine builders follow similar trends in the development of the bore surface finish. However, universal specifications for the plateau surface finish and its parameters are uncommon.

2. MANUFACTURE OF THE BORE SURFACE - MACHINING BEFORE HONING

Before honing, the bore surface is turned in two or more stages using a boring bar fitted with one or more tools. Coarse turning the cast bore surface is followed by fine turning. Coated carbide tools are still used for coarse turning cylinder liners. Fine turning is usually completed with a single ceramic tool.
Cast iron is the major material from which the bore surface is manufactured. Up to 5% of free ferrite, as isolated small grains, may be present within the cast iron matrix. Free ferrite has no effect upon conventional alumina or silicon nitride ceramic tools. However, it greatly impairs the performance and life of cubic boron nitride (CBN) tools. Free ferrite should be practically eliminated if CBN tooling is to be used.

Achieving suitable bore geometry and diameter tolerances before honing is very important. Errors with these cannot be readily corrected during the honing operation. Taper and roundness of the bore presented to the honing machine should be within 25μm. Honing should not be considered as a rectification tool for deficient turning operations. Elimination of sub-surface deformation created by each machining stage must be taken into account when the stock required to produce the honed bore surface from the turned finish is calculated. During the 1970's carbide tools were commonly used for the fine boring operation. Deformation created by carbide fine turning tools extended to between 30μm and 40μm beneath the surface. From Table 1 it can be seen that ceramic fine turning created a greater depth of deformation than carbide tooling.

With ceramic fine turning tools the stock removal on diameter at honing has to be increased to the range 0.10 mm to 0.15 mm to ensure that sub-surface deformation is eliminated.

During honing it is necessary to ensure that the honing stones remain free cutting and do not glaze. Susceptibility to glazing mainly depends on the type of grit and bond hardness of the stone. Initial roughness of the fine turned bore surface and the amount of stock to be removed at honing also effects the tendency for the honing stones to glaze.

To keep the base hone stones free from glazing the Ra of the fine bored surface should be kept within the range of 6μm to 9μm. A fine turned surface smoother than this is unable to keep the base honing stones correctly dressed. Removal of excessive amounts of stock causes the base honing stones to glaze because the honed bore surface becomes too smooth to keep the stones dressed. These difficulties may be overcome by re-selecting the honing stone bond hardness, but this may compromise achieving the desired surface topography. Glazing the honing stones is a major problem. Besides failing to produce the correct distribution of primary valleys of suitable depth, glazed stones increase the depth of deformation at the honed bore surface above what would be expected.

Machining Operation	Depth of Deformation
Coarse machining	70 - 80μm
Fine machining - Ceramic	50 - 60μm
Base Hone - Diamond	25 - 35μm
Base Hone - CBN	10 - 15μm
Base Hone - SiC	10 - 15μm
Plateau Hone - Diamond base	15 - 20μm
Plateau Hone - CBN base	5 - 10μm
Plateau Hone - SiC base	5 - 10μm

Table 1 Machining Operation Versus Deformation

3. TYPES OF HONING STONE

A plateau honed surface comprises of two components as follows:

I. A regular arrangement of deep, oil retaining valleys, created at the base hone stage.
II. Flat load bearing "plateaux", created by the final plateau hone stage.

Diamond honing stones have been used traditionally for either single stage or the base honing stage of a multi stage plateau process, because they give long stone life and consistent cutting characteristics. The diamond grit particles are held within a metallic bond. Table 1 shows that diamond base hone stones create the greatest depth of deformation of all the stone types available.

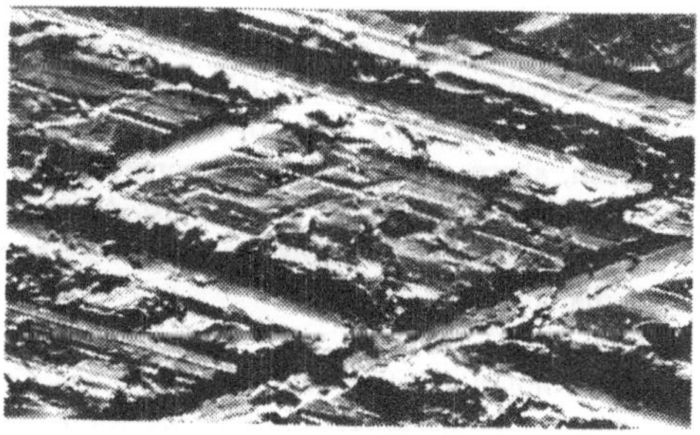

Figure 1 Diamond Base Hone x100

A typical view of a deformed (torn and folded) bore surface after the diamond base hone, but before the final plateau hone stage is shown in Figure 1. Deformation created by the diamond base hones is usually so deep that it cannot be eliminated by the small amount of material removed by the fine grit silicon carbide (SiC) stones used for the final plateau honing stage. Diamond has the most stable crystal structure of all abrasives. As the diamond particles blunt, they do not shear under the influence of high contact pressures. Instead the surface is honed with blunt grit particles that cause deformation. Iron has an affinity for carbon. Although diamond is hard, it is considered that the diamond particles are blunted by diffusion of carbon from the diamond to the cast iron. If it is logistically vital to use stones that have a long life, for example to reduce tool changes, deformation may be reduced by selecting a diamond stone with a softer bond. The softer bond erodes faster causing worn diamond grit particles to fall from the stone face before they have blunted to the point where excessive deformation is created.

Figure 2 CBN Base Hone x100

SiC and CBN stones have a cleaner cutting action, as shown in Figure 2 and Table 1. SiC stones used for base honing generally have a vitreous bond that readily breaks down. High stone wear caused by this continuously exposes fresh SiC grit particles at the abrading surface giving a clean cutting action. High stone wear found with SiC honing stones does not usually incur a cost penalty, because they are typically less than 5% the cost of diamond or CBN stones. The more frequent number of tool changes to replace worn SiC base honing stones can prove inconvenient when manufacturing long production runs of high volume items.

CBN stones have a metallic bond like diamond. Unlike diamond the CBN crystal structure has certain planes of weakness. When a CBN particle becomes blunt, the forces within it increase to the point where the shear stress of the crystal is exceeded and the CBN crystal fractures to reveal fresh, sharp cutting facets. Compared with diamond this causes higher grit consumption and the bond hardness has to be adjusted to ensure that CBN particles are readily exposed. The cost of diamond and CBN stones is very similar, but CBN stones have a much shorter life than diamond stones.

Except for certain high volume gasoline engine applications, fine grit SiC stones are used for the final "plateau" stage of a plateau honing process. Stone wear during the plateau stage is much lower than during the base hone because of the small amount of stock removal (perhaps only 20μm on diameter) required to form the final surface. Therefore, only a few plateau strokes are usually needed.

4. SURFACE APPEARANCE AND PROPERTIES

In the early days of plateau surface technology many drawings demanded only Ra (typically 0.6µm to 1 .2µm) and sometimes Rt, often with a statement that the bore must be plateau honed. This led to a wide variation of surfaces because the Ra could be achieved by having either very deep valleys with smooth plateaux or shallower valleys and rough plateaux. Traces that illustrated satisfactory and unacceptable surface profiles were sometimes included to control the wide variation caused by the inadequacy of parameters available.

Meaningful measurement of the plateau only became available by graphing the Abbott-Firestone curve. Some early work at GKN involved measuring the change in cylinder bore surface finish during the life of a diesel engine cylinder liner from new. Wear of the bore surface by the piston ring during bedding in and early life basically completed the task of forming the load bearing plateaux that were later created by the final plateau honing stage. Wear was quite rapid until the cylinder bore surface had been worn, to give a bearing area of about *50%* to 70% to the piston ring. Early GKN plateau honed surfaces aimed to create this bearing area within a few microns of the highest peak on the bore surface. Macro bearing area (TPA) at a single slice level, for example 2µm was used to control the plateau area. The depth selected was based upon what was considered the level of controlled wear acceptable during bedding in and to provide a surface finish trace that possessed the plateaux desired.

Figure 3 Early Plateau Surface x100

Increased combustion pressures led to higher ring pressures, during the 1970's and 1980's. This made adhesive wear or scuffing resistance an important factor. Asperities and peaks removed from the base honed surface by the final plateau honing stage significantly improved scuff resistance of the bore surface. The mechanism of scuffing was considered to involve the interaction of asperities on the bore surface and piston assembly, which punctured the lubricant film and permitted adhesive wear. Plateau surfaces gave reduced oil consumption because the oil volume retained by the surface was lower than that of a traditional cross honed surface. Piston top land deposits caused by burned oil were also decreased by the reduced oil consumption. Therefore, besides scuff resistance and reduced oil consumption, reduced bore polishing was also seen in some engines.

Compared with plateau honed surfaces created for present and future engines, surfaces from the 1980's had rough plateau tops and primary oil retaining valleys of inconsistent depth as shown in Figure 3. Often the plateaux had a significant oil reserve of their own. The driving force throughout the 1990's has been to control the distribution and depth of primary valleys to a greater extent and create smoother, flatter plateaux to reduce oil retention and therefore consumption further.

Greater control has been applied to the honing operation and in particular to the base hone stage. Efficient filtration of the honing fluid and control of its cutting characteristics is necessary to ensure that the base hone stones cut cleanly and consistently. Honing debris in the fluid create scores and spurious valleys at the bore surface. Cutting action is lost if the honing fluid blend is incorrect. This often gives increased surface deformation.

A typical plateau bore surface for a Euro III. diesel engine is shown in Figure 4. From this it can be seen that for the 1990's the primary oil retaining valleys are narrower and shallower than those of the early 1980's surface shown in Figure 3

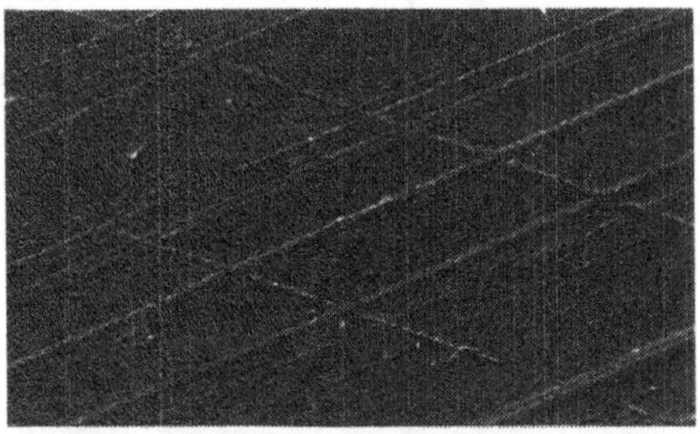

Figure 4 1990's Plateau Surface x100

Control of the primary valleys is often the most difficult part of the honing process. Smaller grit SiC plateau stones are generally used in the 1990's when compared with the 1980's. This enables the creation of smoother plateaux, that have a lower oil reserve. Smoother, flatter plateaux reduce the lubricant film thickness.

Greater emphasis has been placed upon minimising deformation at the bore surface to reveal more of the graphite flakes, which impart adhesive wear resistance. Nowadays many specifications include statements or sections that identify deformation or exposed graphite requirements. Reduction in the depth of the primary valleys has caused some concern about the tolerance of modern plateau surfaces to bore wear. With some surfaces only about 5~m wear per side can be tolerated before the majority of primary valleys are lost. This coupled with concerns about abrasive wear caused by the recirculation of hard PM with the introduction of EGR, has led to renewed interest in induction hardened cylinder bores to resist abrasive wear of the finer surface. 1990's diesel plateau surface technology has been transferred recently to a gasoline engine that suffered cold piston scuffing. Piston ring wear has been reduced and cold piston scuffing has all but disappeared after the revised surface finish had been introduced.

5. SPECIFICATION OF THE SURFACE FINISH

Surfaces comprise of three characteristics:

i. **Form.** This is the design of the surface's shape, for example the cylinder bore diameter.
ii. **Waviness** or undulations are superimposed upon the form and are normally caused by the machining operations, for example fine turning of the bore surface.
iii. **Roughness** is superimposed on the waviness. There is no strict definition to separate roughness and waviness. A cut-off point is chosen to separate the two.

Measurement conditions, for example filter type and cut off length must be established before parameters to define the surface finish are selected. A filter cut-off length (λc or Lc) is used to separate roughness and waviness. Years ago cut off was chosen according to the Ra of the surface with little regard to peak spacing. Plateau surfaces have a periodic arrangement of oil retaining valleys with a spacing of about *0.5* mm. To ensure these are included in the calculation of roughness parameters the cut-off should be at least 2.5 times the valley spacing. The cut off is set at discrete values, for example 0.25 mm, 0.8 mm, 2.5 mm, etc. A 2.5 mm cut off is the most suitable value from this range to capture the information required. This cut off will normally require an evaluation length (Ln) of 12.5 mm (ie 5 cut-off values), plus an additional cut-off length for filtration purposes.

Two basic types of filter, analogue (2RC) and phase corrected (PC or Gaussian) are available. Filters distort the profile. A 2RC filter will create little distortion of a sinusoidal profile. Plateau surfaces are far from sinusoidal and the 2RC filter causes much overshooting of the filtered profile. The PC filter uses a Gaussian centre weighting flinction that eliminates horizontal phase shift, giving minimal vertical distortion to the profile mean line. This creates far less distortion than the 2RC filter. Therefore, the PC filter is recommended for use when measuring plateau honed surfaces.

Philosophies for the specification of plateau surfaces fall into three main categories.

I. Conventional Parameters. Mean peak to valley height (Rz-DIN) and normalized peak count (Pc) are used to measure primary valley depth and frequency. Rz-DIN is the mean of the maximum peak to valley height for each of the five cut off lengths that make up the evaluation length. Rz-ISO should not be used because this calculates the mean distance between the five highest peaks and five deepest valleys irrespective of where they occur within the evaluation length. Pc is usually set at a distance below the plateau height, for example 3 μm below the mean line. Plateau area is measured using the unfiltered macro bearing area (Tpa). The piston assembly will experience both roughness and waviness as it travels along the cylinder bore. Therefore, the unfiltered parameter Tpa is preferred to the filtered micro bearing area Tpi. A reference line typically between 2% and 5% is included in calculation of the Tp value to give greater stability to this parameter, otherwise measurement would be affected by an isolated, spurious peak.

A reliable measure of the plateaux can be obtained by measuring the bearing area at one level, for example 2μm. Specification values for plateau surfaces are shown in Table 2.

Parameter	1980's	1990's
Rz-DIN (μm)	5-10	4-8
Pc at -3μm (cm^{-1})	10-20	15-20
Tpa at 2μm with 2% ref (%)	70-90	80-90

Table 2 Conventional Surface Finish Parameters

From Table 2, it can be seen that while ranges set in the 1980's are similar to those for the 1990's the primary valley depth has reduced and the plateau area has moved to the upper half of the 1980's specification. Tighter limits for the 1990's have demanded greater control of the honing operation. The effects of these changes are also seen by comparing Figures 3 and 4.

II. Custom Tpa Bands. Another approach divides the Abbott-Firestone curve into discrete regions to enumerate the plateau top roughness removed during bedding in, the functioning area of the plateau and the oil retaining primary valleys. The Abbott-Firestone curve gives no measure of the frequency or spacial arrangement of valleys. It may be prudent to include Pc as described for the conventional parameters. Typical ranges for Euro I and II surfaces are given in Table 3. In this example the top and oil parameters include 2% reference levels at either end of the curve to give stability to the results. Changes from Euro I to Euro II include reducing the primary valley depth and the plateau depth to achieve a particular bearing area range. This means that the plateaux are smoother and flatter for Euro II.

Parameter	Euro I Range	Euro II Range
Top (Tpa at 20% - 2%)	0 - 1.5μm	0 - 0.7μm
Core (Tpa at 70% - 20%)	0.3 - 2.5μm	0.3 - 1.5μm
Oil (Tpa at 98% - 70%)	2.0 - 7.0μm	2.0 - 4.0μm

Table 3 Tpa Bands For Euro I and II Plateau Surfaces

III. Rk Parameters Parameters specified in DIN 4776 are based upon the Abbott-Firestone curve. Due to their method of calculation from the curve they are not as straightforward to understand as the custom Tpa band approach. This has possibly restricted their use. From Table 4 it can be seen that changes for the 1990's included:

i. Reduction of Mrl, reduced peak height and core profile, which meant making the plateaux smoother and straighter than in the 1980's.

ii. Increased consistency of the primary valley depth demanded by the smaller reduced valley depth range. Smoother, straighter plateaux would also emphasise the arrangement of primary valleys.

Parameter	**1980's**	**1990's**
Metal Ratio Mr1 (%)	8 max	5 max
Metal Ratio Mr2 (%)	75-85	70-85
Reduced Peak Height (μm)	0.8 max	0.3 max
Reduced Valley Depth (μm)	1.0 - 3.0	2.2 - 3.0
Core Profile (μm)	1.0 - 3.0	0.5 - 1.5

Table 4 DIN4776 Parameter Ranges

6. CONCLUSIONS

Honing technology and surface finish measurement techniques have evolved markedly over the past 25 years to give reduced oil consumption, scuff resistance by permitting modification and refinement of the bore wearing surface with improved consistency, and reliable measurement. There is consensus driving towards clean cut surfaces with smooth, flat plateaux and a regular, consistent arrangement of primary oil retaining valleys. However, different philosophies have developed to specify the surface. Future development is expected to focus on the honing process to improve consistency of the surface finish further. A trend towards the use of even smoother plateaux may continue. Further reduction of primary valley depth to decrease oil consumption may demand the adoption of hardened treatments to overcome the problem of destruction of the surface's low oil reserve after very limited "normal" bore wear.

The development of high-strength aluminium tin silicon copper bearing alloys for automotive applications

W P BROWN
Glacier Vandervell Europe, Kilmarnock, UK

1) INTRODUCTION

The design characteristics of modern engines place increasing demands on the crankshaft system comprising the bearings, bearing housings, crankshaft, connecting rods and oil. These characteristics usually result in increased specific loading on the bearings, due to either the reduction in size of the components, the increase in applied loading or both.

The highest fatigue strength requirements use a trimetal bearing based on an overlay plated copper-lead substrate which is bonded to a steel backing. The slightly lower strength requirements use a bimetal system based on aluminium alloys bonded to a steel backing. These do not require an overlay and are therefore cheaper to manufacture. These materials have suitable bearing properties. The challenge for bearing manufacturers is to develop these materials for use in ever more demanding applications.

Over many years, Glacier has developed aluminium tin silicon copper alloys as improvements to the original aluminium tin copper bearing alloy. At the present time, the new alloys are being used in automotive applications. The aluminium tin silicon alloys have been found to be capable of further development to improve their performance. The nominal compositions of all thc alloys are shown in Table 1.

Alloy Type	Al (Wt %)	Sn (Wt %)	Si (Wt %)	Cu (Wt %)
AlSnCu	Bal	20.0	-	1.0
AlSnSiCu (original)	Bal	12.0	4.0	2.0
AlSnSiCu (improved)	Bal	12.0	4.0	1.0

Table 1 Composition of Glacier Aluminium Based Bearing Alloys

2) BEARING ALLOY DEVELOPMENT

The main properties necessary for successful bearing performance are fatigue strength to withstand the cyclic loads, conformability to allow the bearing to accommodate slight misalignments and irregularities in the bearing system and compatibility to reduce the seizure risk.

The first generation of aluminium tin silicon alloy involves utilising the copper addition to provide maximum strength to the aluminium matrix. The further improvements to the system involve optimising the copper composition to maintain sufficient strength while also obtaining improved ductility of the aluminium matrix. Fig 1 shows how the tensile properties of the AlSnSi alloys vary from room temperature to 190 C, typical of the upper range of running temperatures observed in modern engines.

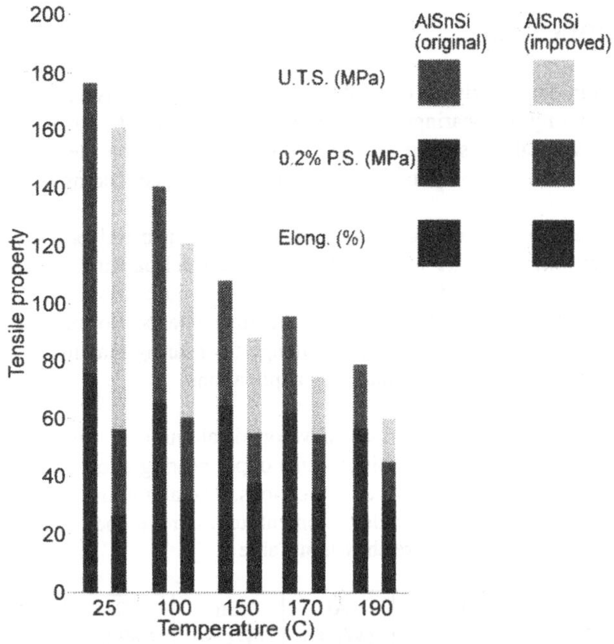

Fig 1 Variation of Mechanical Properties with Temperature for the AlSnSi Alloys

Corresponding hardness values are shown in Fig 2.

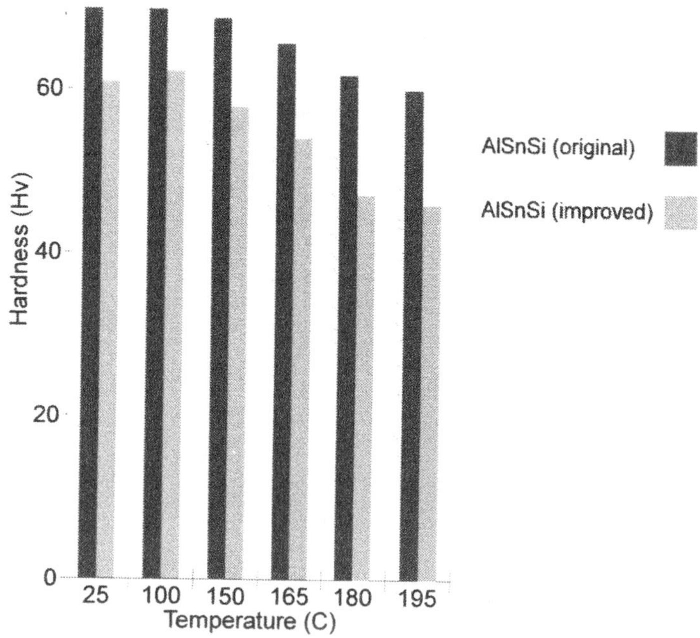

Fig 2 Variation of Hardness with Temperature for the AlSnSi Alloys

However, the consideration of these simple mechanical properties does not give sufficient information on whether a particular alloy will perform as a good bearing material. For this reason, the combination of strength and ductility, known as the 'Toughness Index'[(1)] has been used to assess material characteristics, represented by the formula:

$$\text{Toughness} = 2/3(\text{UTS} * \text{Total Elongation})$$

This approximates to the area under the stress/strain curve, Fig 3.

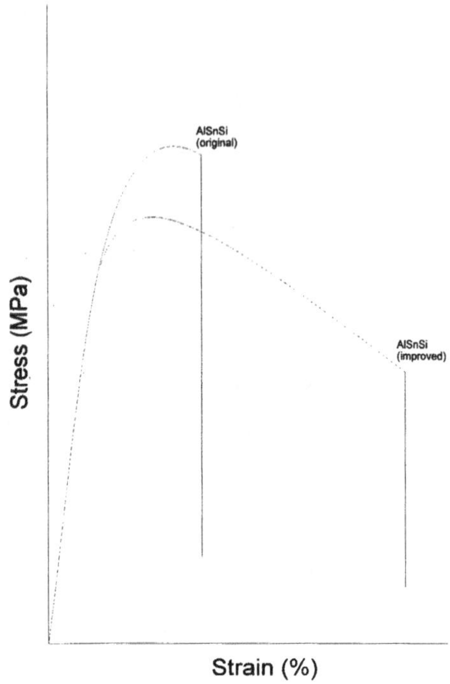

Fig 3 The Toughness Index Principle

This is an indication of the ability of a material to absorb energy in the plastic range. A material which may be of slightly lower strength but higher ductility may be able to deform to a greater extent under the same loading conditions, therefore reducing the subsequent loadings on the lining.

The implication of this is that for a high strength bearing, the alloy should not only possess sufficient high temperature tensile strength, but must also possess an adequate high temperature ductility to ensure sufficient conformability. Fig 4 shows the toughness with temperature for the alloys.

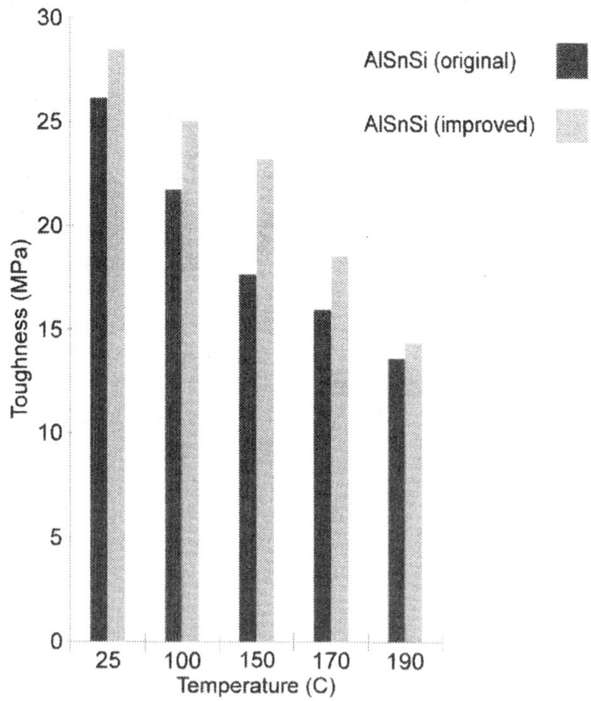

Fig 4 Variation of Toughness with Temperature for the AlSnSi Alloys

The concept of toughness is strengthened by measurements of bearing fatigue strength in a 'Sapphire' bearing test machine, Fig 5.

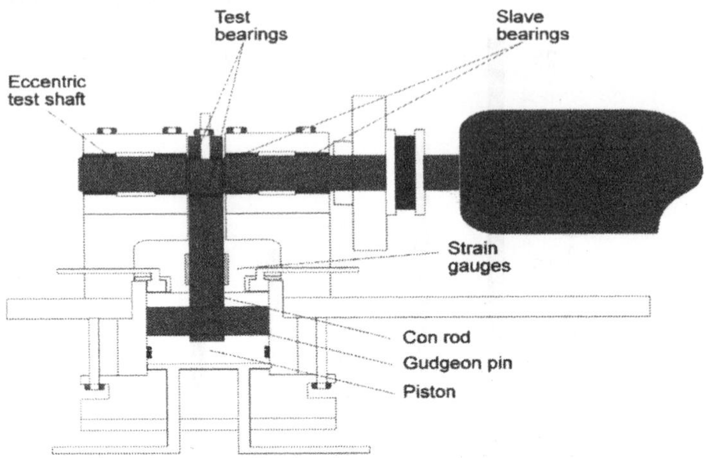

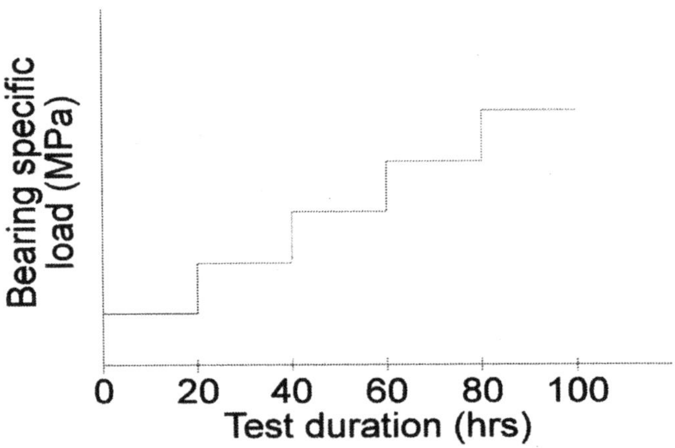

Fig 5 The Sapphire Test Machine with Load Ramp Profile for Bearing Fatigue Test

Fig 6 shows the fatigue strength as measured by the Sapphire test for the alloys. In spite of the slightly reduced tensile strength of the improved alloy over the earlier one, the bearing fatigue strength is unchanged.

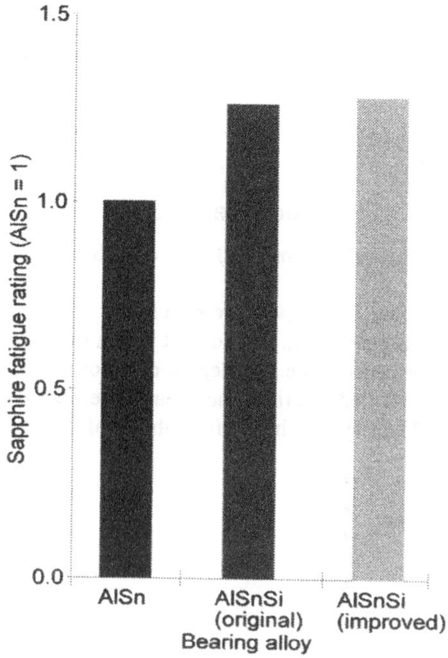

Fig 6 Sapphire Fatigue Strength of the AlSnSi Alloys Relative to AlSn

It has also been shown that a Sapphire machine can be used to determine a conformability result for bearing alloys by steadily ramping up the load while running the bearing in a thin oil film regime, Fig 7.

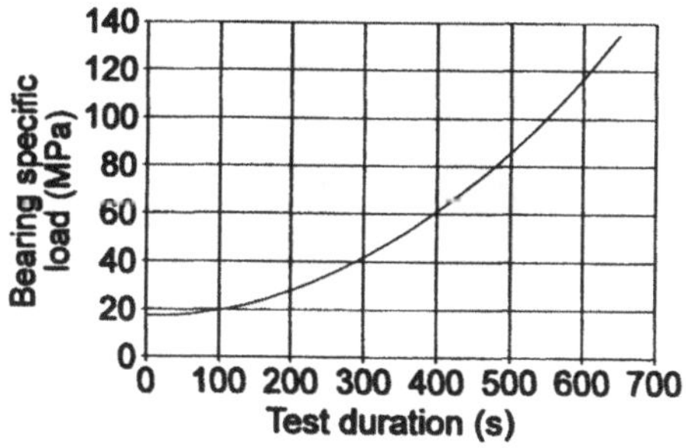

Fig 7 Load Ramp Profile for Bearing Conformability Test on Sapphire Machine

In this test, the bearing temperature is measured while the load is increased, Fig 8.

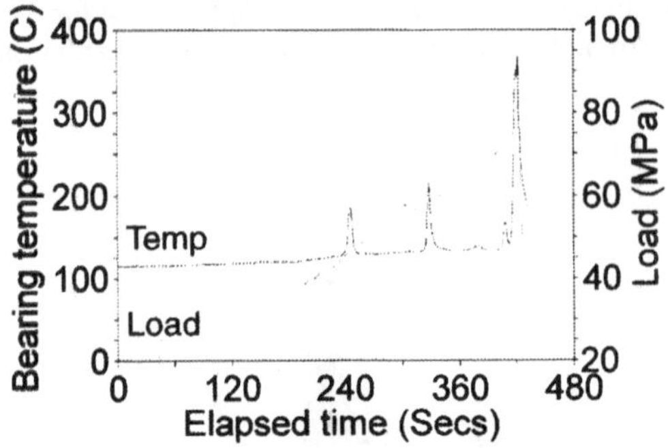

Fig 8 A Typical Conformability Test for the Improved AlSnSi Alloy

The trace shows the occurrence of temperature excursions which then self correct as the load increases (oil film thickness decreases). The load at which the first excursion is observed is taken as a measure of conformability, i.e. the degree of deformation of the material under the applied load. The greater the deformation, the greater the applied load before the first temperature excursion occurs. Fig 9 shows the conformability of the various alloys.

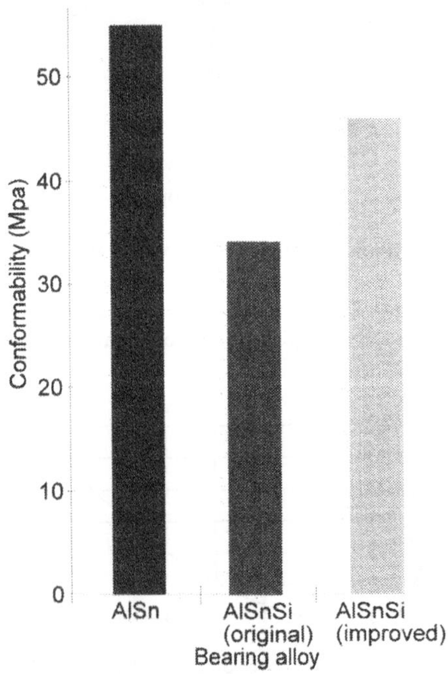

Fig 9 Sapphire Conformability Test Results

The original aluminium tin alloy, shows very high conformability compared to the aluminium tin silicon alloys. However, the improved alloy shows a useful increase over the earlier one.

The result shown has been further verified by comparing the performance of the earlier and improved versions in a modern four cylinder, 16 valve 2.0 litre petrol engine. Following a short 20 min run in at 850 rpm, then a four hour full speed (6500 rpm) full load test, the bearings appeared as shown in Fig 10.

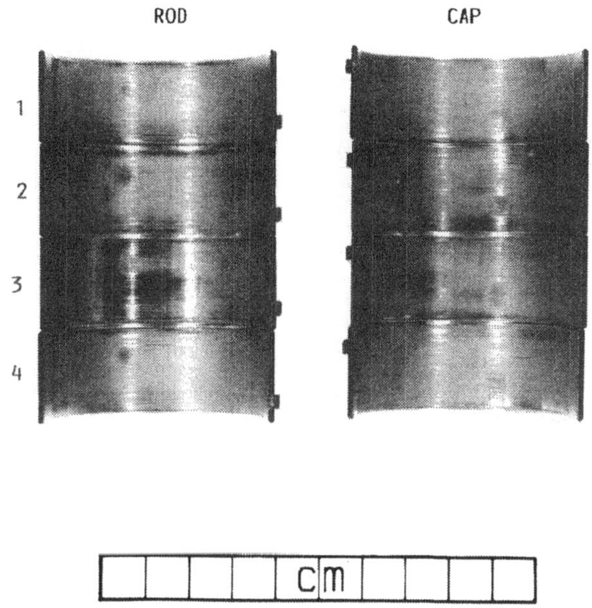

Fig 10 (a) The Surface Appearance of the Original AlSnSi Alloy Following the Engine Test Programme.

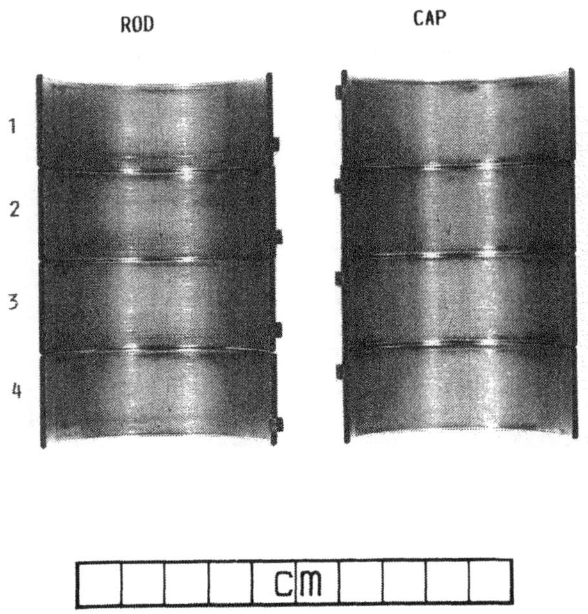

Fig 10 (b) The Surface Appearance of the Improved AlSnSi Alloy Following the Engine Test Programme.

Although both sets of bearings were successful in operation, the bearings made from the original alloy have a greater degree of surface polishing. This is consistent with the lower conformability measured by the temperature excursion method.

Under the most extreme operating conditions, when the material can no longer conform or condition the shaft sufficiently to avoid thermal runaway, i.e. no recovery of the temperature excursions in Fig 9, seizure will take place. However, reducing seizure involves matching all the components of the bearing system and this will be described in more detail in the next discussion.

Other properties which are important for successful bearing operation are corrosion resistance, wear resistance and cavitation erosion resistance. Corrosion resistance is low for both types of aluminium alloy compared to trimetal systems, while the aluminium tin silicon alloys show useful gains in wear and cavitation erosion resistance compared to the aluminium tin alloy.

3) CURRENT STATUS

The original aluminium tin silicon bearing alloy is in volume production and is supplied successfully for a number of applications up to 55 MPa specific load. The improved alloy is also, since May 1995, being supplied into a production application. In addition, it is under test in several other applications and is giving successful results, especially where the system is sensitised to geometric imperfections through thin oil films.

4) FURTHER DEVELOPMENTS

The next stage of improving the aluminium tin silicon alloys is to concentrate the studies into the interfacial region between the alloy and the steel. At present, this region consists of a thin pure aluminium layer. There is evidence which suggests that overall bearing strength could be increased without any reduction in the surface properties by replacing this pure layer with a stronger non tin containing aluminium alloy. Early results have verified this.

REFERENCE

1) G E Dieter, 'Mechanical Metallurgy', McGraw-Hill Book Company, 1988, Pg. 283

Developments in camshaft and follower technology

A C BROOKS FIDiagE
Diesel Engine Engineering, Ford Motor Company Limited

INTRODUCTION

Internal combustion engine valve train design has changed significantly over recent years, moving from the "traditional" 2-valve per cylinder push-rod designs to the current range of multi-valve overhead camshaft designs. This article gives an overview of the most common designs, and the camshaft and follower technologies and materials associated with them. It also examines some of the recent developments towards weight, friction and cost reduction associated with the drive for improved engine efficiency and reduced emissions at affordable cost.

OHV PUSH-ROD VALVE TRAIN :

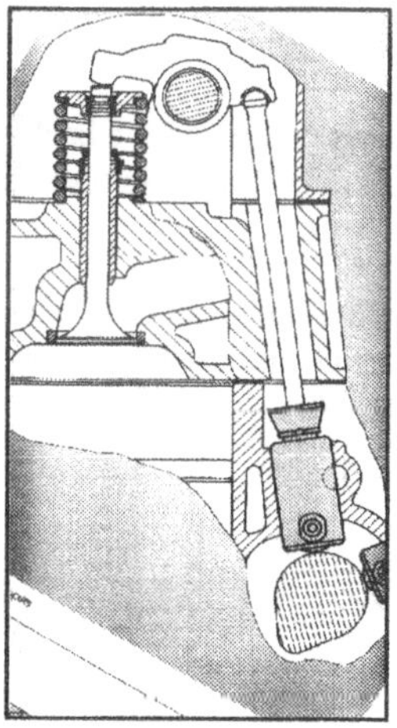

Fig. 1

The illustration in Fig 1. shows a typical push-rod operated overhead-valve layout. The tappet can be either sliding or rolling contact, with mechanical or hydraulic lash adjustment. The Advantages and Disadvantages of this layout are :

- ➤ Advantages:
    - ➤Simple Camshaft Drive - Gear, Short Chain or Belt.
    - ➤ Low Engine Height.
    - ➤ Simpler Head Design.

- Single Camshaft Possible for V-Engines - Fig. 1, in fact, shows one half of a V-Engine with a Roller Follower and a Hydraulic Lash Adjuster (HLA) built into the tappet.
- Disadvantages:
  - High Reciprocating Mass.
  - Low Valve Train Stiffness.

Fig. 2

This layout is still used in a number of heavy truck and large marine and industrial engines.

The illustration in Fig 2. shows its use, with a bridge piece, in the Mercedes 500 Series 2.0 L / Cyl Direct Injection Diesel Engine Family, to operate 4 valves per cylinder with skewed valve layout for optimum port flow design.

1997 European Ford engines still using an OHV push-rod valve train are the Transit 2.5 litre DI (Direct Injection) Diesel Engine and the Fiesta and Ka 1.3 litre "Endura" (also known as the HCS or Valencia) gasoline engine.

WHY OVERHEAD CAMSHAFT ? :

The move to overhead camshaft engines has been primarily driven by the move, in Europe and Japan, to smaller, higher speed, engines. A large number of American V-engines still use push-rod valve trains.
Valve Train Mass and Stiffness become increasingly significant as engine speeds increase, and Overhead Camshaft (OHC) Valve Trains are inherently lighter and stiffer than Push-Rod designs.

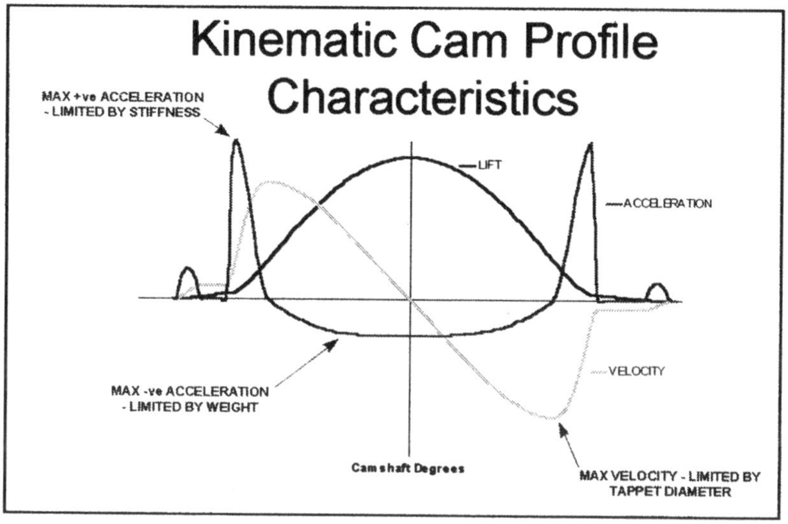

Fig. 3.

The significance of Valve Train Stiffness and Mass can be seen from the Kinematic Cam Profile characteristics shown in Fig 3.

The maximum Initial Acceleration which can be generated as the valve starts to open is limited by the Valve Train Stiffness. If it has to be held to a low value then it is difficult to achieve the required maximum Valve Lift, and insufficient air gets into the cylinder to achieve the Performance and Emission Targets.

The maximum "Negative Acceleration" in the area around the Cam Nose is limited by the resultant of the Valve Train Mass trying to separate the system and the Valve Spring Load trying to hold it together. Again, this constrains the maximum Valve Lift which can be achieved.

The maximum Velocity is proportional to the "Eccentricity" of the Cam-to-Tappet contact point on the Tappet. Therefore Tappet Diameter constrains maximum Velocity, which in turn limits the maximum Valve Lift which can be achieved. However a larger Diameter Tappet increases Valve Train Mass, which in turn affects the maximum "Negative Acceleration" limit, so it can be seen that Cam Profile Development is a "Trade-Off" of several conflicting parameters.

OVERHEAD CAMSHAFT CONFIGURATIONS :

Some typical Overhead Camshaft Valve Train configurations are shown in the following illustrations :

OHC FINGER FOLLOWER VALVE TRAIN :

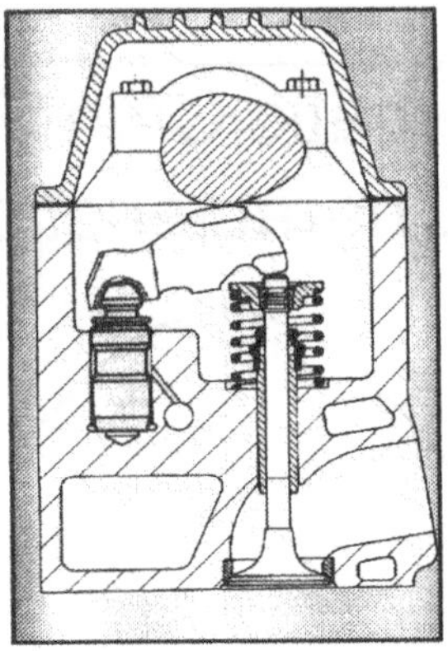

Fig. 4

The illustration in Fig 4. shows an OHC Finger-Follower valve train layout. It can be used with either a mechanical or hydraulic end pivot lash adjuster. The illustration shows hydraulic lash adjustment.

The Advantages and Disadvantages of this layout are :

- Advantages:
  - Low Reciprocating Mass.
  - High Stiffness.
- Disadvantages:
  - High Friction.
  - Pure Sliding Cam-to-Follower Interface Requires Positive Lubrication and is Prone to Wear.

This layout was used in the well-known Ford "T88" sohc petrol engine from the early 1970's until the early 1990's, and in a similar GM engine of the same era. Anyone who has owned one will testify to the design being prone to wear and dependent on the oil supply rail jets being kept clear. The design kept aftermarket camshaft and follower suppliers in business for years.

CVH OHC VALVE TRAIN :

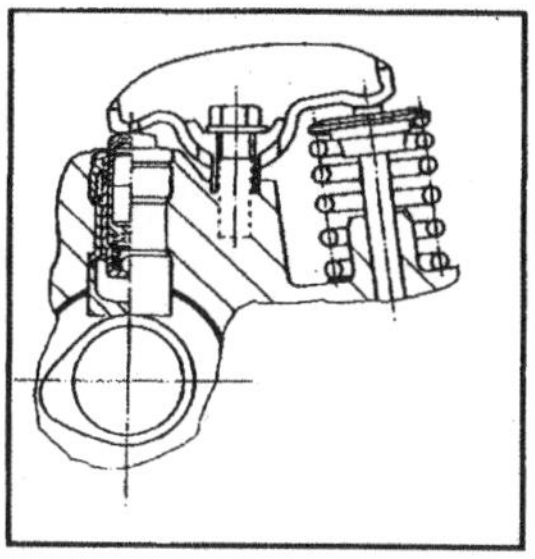

Fig. 5

Fig 5. Shows a "CVH" (Compound Valve Head angle) Valve Train layout. This is effectively a push-rod layout with the push-rod removed and the camshaft moved up into the cylinder head. The use of a spherical cup, instead of a shaft, for mounting the rocker arm allows the valves to be inclined axially as well as laterally for improved combustion - hence the name CVH. The picture shows a sliding follower with hydraulic lash adjuster, typically used in Europe. American versions of this layout tend to use a roller follower with hydraulic lash adjuster, while high performance derivatives often use mechanical lash adjustment for increased stiffness.

The Advantages and Disadvantages of this layout are :

- Advantages:
  - Reduced Reciprocating Mass.
  - Increased Valve Train Stiffness.
- Disadvantages:
  - Small Tappet Diameter Constrains Cam Profile Design.

TYPICAL MODERN DOHC VALVE TRAIN CONFIGURATIONS :

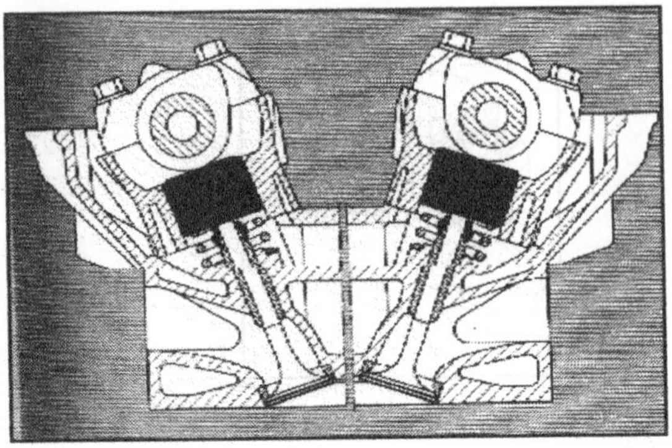

Fig. 6
Direct-Acting Bucket Tappet

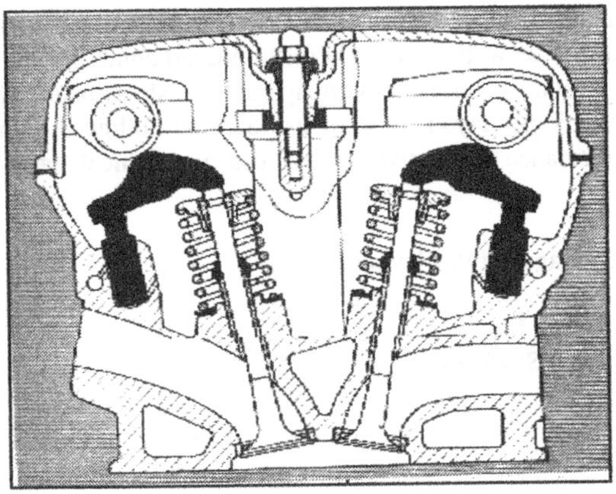

Fig. 7
Roller Finger Follower

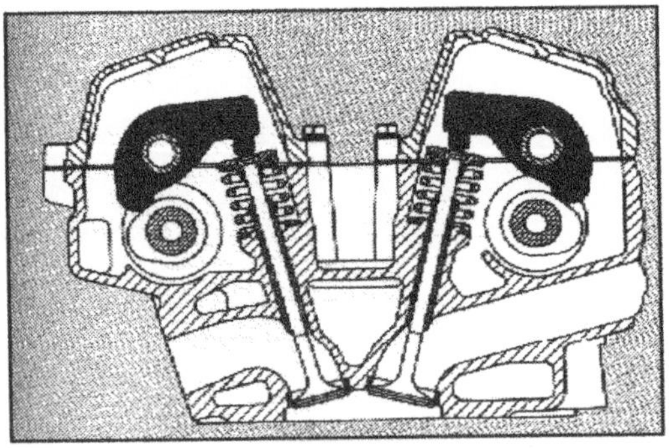

Fig. 8
Roller Rocker Arm

Figures 6 to 8 show three of the most common modern DOHC valve train configurations.

All three configurations can be used for 2 or 4 valve per cylinder designs, and can also be used for single camshaft 2 valve per cylinder designs.

The finger follower and rocker arm layouts can also be used for single camshaft 3 valve per cylinder designs.

Valve lash adjustment can be hydraulic or mechanical (the illustrations used show hydraulic lash adjustment).

The layouts shown are typical of DOHC gasoline engines with inclined valves to give a "pent-roof" combustion chamber. Diesel engines require vertical valves for optimum combustion which makes the dohc direct acting bucket tappet design infeasible because the camshafts are too close together. The dohc bucket tappet layout can be used if the valves are inclined very slightly (typically at about a 5° included angle). It is then possible to drive one camshaft with a belt or chain and drive the other one from it using gears.

Further advantages and disadvantages of the various layouts are summarised below:

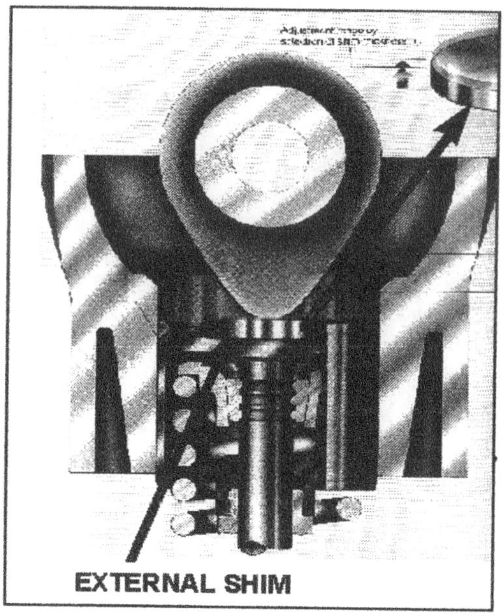

Fig. 9
Direct Acting Bucket Tappet.
Mechanical Lash Adjustment
with External Shim

DIRECT-ACTING - EXTERNAL SHIM :

- Advantages:
    - High system stiffness.
    - Optimum material choice can be used for the shim. A commonly used material is phosphated 100Cr6 Bearing Steel, through-hardened to around 62 HRC. In European and Japanese engines this is usually run against a Chilled Cast Iron cam lobe with a hardness of around 50 HRC.
- Disadvantages:
    - Requires adjustment on initial engine build.
    - Effective cam contact diameter is less than tappet diameter, unless an expensive "stepped" shim is used.
    - Service check and adjustment is required.

DIRECT-ACTING - INTERNAL SHIM :

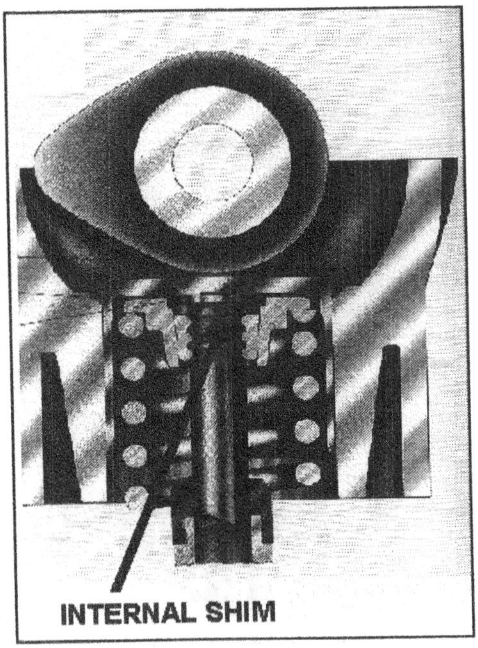

Fig. 10
Direct-Acting Bucket Tappet
Mechanical Lash Adjustment
with Internal Shim

➢ Advantages:
- ➢ High system stiffness.
- ➢ The full tappet diameter is available for cam contact, which increases cam profile design flexibility.
- ➢ Optimum shim material can be used.

➢ Disadvantages:
- ➢ Requires adjustment on initial engine build.
- ➢ Service check and adjustment is required, and camshaft has to be removed for service adjustment.
- ➢ Limited material choice for the tappet face. This has to be the same as the tappet body unless a separate top plate is welded in with a consequent cost penalty. A commonly used tappet body material is 16MnCr5 steel. This can be hardened to around 600 HV5 (55 HRC), but doesn't have the wear resistance of 100Cr6.

DIRECT-ACTING - HYDRAULIC LASH ADJUSTER :

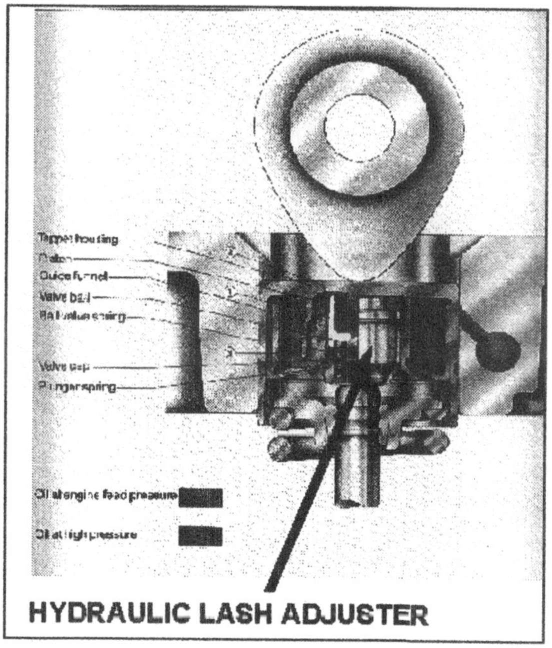

Fig. 11
Direct Acting Bucket Tappet
Hydraulic Lash Adjustment

➢ Advantages:
- ➢ No lash adjustment on initial engine build.
- ➢ The full tappet diameter is available for cam contact.

- Compensates for valve seat wear in service - up to a point. The hydraulic element only has a finite operating range, and the majority of this range can easily be used up by the geometrical tolerances of the cylinder head and the valve train components.
- Disadvantages:
  - Not as stiff as mechanical adjustment. The hydraulic element reduces system stiffness due to the compression of any entrained air in the oil and the "leak-down" sinkage, which is designed to occur as the valve opens to guarantee complete closing.
  - Limited material choice for the tappet face. The same limitations apply as for internal shim mechanical adjustment.

ROLLER FINGER FOLLOWER :

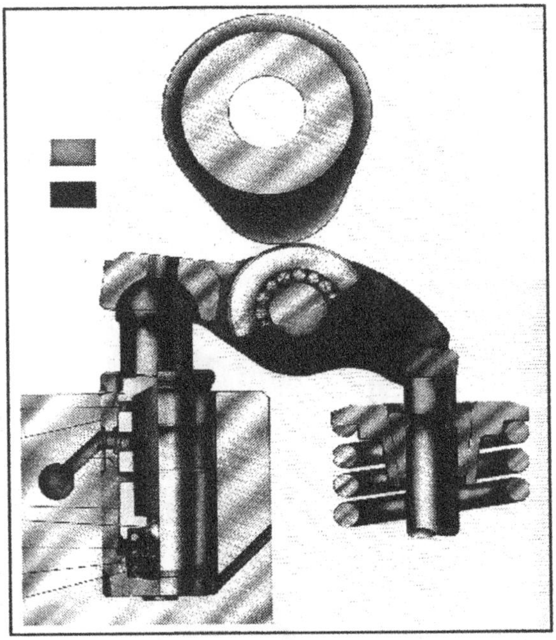

Fig. 12
Roller Finger Follower with
End-Pivot Lash Adjuster

- Advantages:
  - High Stiffness (Although not as stiff as direct-acting design).
  - Roller follower gives much lower friction than sliding contact design (Fig.4) and doesn't require positive lubrication.
  - A low-cost pressed-steel follower can be used, with minimal machining requirement.
- Disadvantages:
  - Roller follower creates higher contact stresses than sliding follower and is likely to require a "re-entrant" cam form and more expensive cam lobe material or treatment.

ROLLER ROCKER ARM :

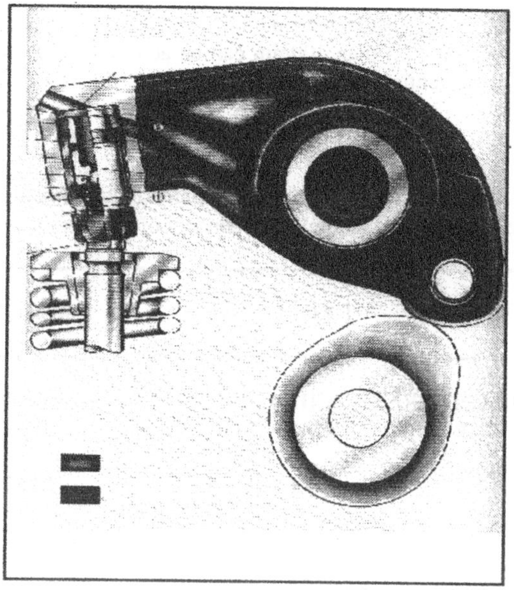

Fig. 13
Roller Rocker Arm

- Advantages:
  - Low engine height compared to direct acting and finger follower designs.
  - Low friction.
- Disadvantages:
  - Low stiffness compared with direct-acting or finger follower designs. This is due primarily to the stiffness of the rocker shaft.
  - Roller follower disadvantages, as for the finger follower design.
  - Die cast aluminium rocker arm requires much more machining than pressed steel finger follower.

All Roller Follower designs have an inherent disadvantage over Sliding Contact designs in that the Hertzian contact stress between two curved bodies (the cam lobe and the roller) in a roller follower design is significantly greater than that between one curved body (the cam lobe) and one flat, or much less curved, body (the tappet or finger, etc.) in a sliding contact design.

The high contact stress levels associated with roller followers mean that they have historically been run against steel cam lobes which can operate at 2 to 3 times the stress levels which have traditionally been accepted for cast iron.

Even with a steel cam lobe it is necessary to have a larger nose radius for a roller follower than a sliding one, and this means that to generate the same lift within the same cam angle it is often necessary to have a "re-entrant" (concave) radius on the cam flanks. This is more difficult to grind than a convex radius, and the degree of re-entry is limited by the diameter of the grinding wheel used - a conventional, oxide, grinding wheel is 600mm in diameter, whereas a "new technology" CBN (Cubic Boron Nitride) wheel can go down to 400mm diameter. This allows the concave radius to go down from approximately 400mm to 300mm. Another recent innovation is Belt Grinding which uses an abrasive belt passing over a form blade. Re-entrant radii of less than 100mm can be achieved using this process, but it is very slow compared with conventional or CBN grinding.

MATERIALS :

Table 1 below shows some typical cam and follower material combinations, and contact stress capability. (Contact stress capability is largely independent of valve train geometry).

	CAM LOBE	FOLLOWER	MAX HERTZ STRESS
SLIDING CONTACT	CHILLED GREY CAST IRON	CHILLED GREY CAST IRON	700 MPa (100,000 psi)
		100Cr6 STEEL	900 MPa (130,000 psi)
ROLLING CONTACT	FORGED 100Cr6 STEEL	100Cr6 STEEL	1800 MPa (260,000 psi)
	SINTERED STEEL	100Cr6 STEEL	1000-1500 MPa (145-215,000 psi)
	AUSTEMPERED DUCTILE IRON	100Cr6 STEEL	1500 MPa (215,000 psi)
	CHILLED GREY CAST IRON	100Cr6 STEEL	1100 MPa (160,000 psi)

Table 1

This is not an exhaustive list of the material combinations available to the engine designer. Hardenable iron, for instance, can be used instead of chilled iron for sliding contact applications, particularly in countries where there is no facility for producing chilled iron. The Hertz stress figures quoted are only intended to be a general design guide, and the maximum contact stress for a given application will depend on the precise geometry and lubrication characteristics, etc. of that application.

FUTURE DEVELOPMENTS :

The current drive in Europe and Japan is towards lighter, lower friction, more fuel-efficient engines. This is fuelled to a large extent by the increasingly tighter exhaust emission legislation being planned for the next millennium.

This has led to recent developments in valve train system and component designs aimed at reducing engine weight, friction and cost.

Some technologies for weight, friction and cost reduction currently in production, or under development are :

- Hollow-Cast Camshafts.
- Assembled Camshafts.
- Lightweight Tappet (and Valve) Components.
- Operation of 2 Valves by 1 Cam Lobe.
- 4 Valves per Cylinder Operated by a Single Camshaft.

HOLLOW-CAST CAMSHAFTS :

Hollow-cast camshafts can be used either purely for weight saving, or to replace an expensive gun-drilling operation when a hollow shaft is required to transfer oil between bearing journals. Recent developments in profiled cores can give hollow-cast camshafts competitive weights to assembled shafts. Fig 14, below, illustrates three different applications of hollow-cast camshaft technology, and the typical weight savings which can be achieved with them :

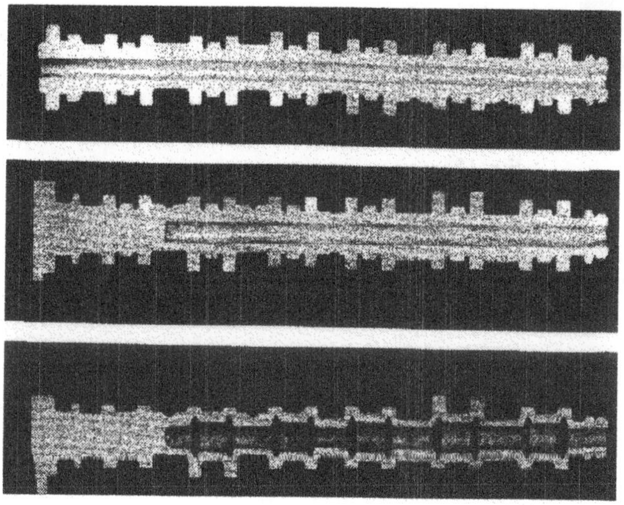

- Straight core with through hole. 15% weight reduction versus solid shaft.
- Straight core with blind hole. 11% weight reduction versus solid shaft.
- Profiled Core with blind hole. 22% weight reduction versus solid shaft.

Fig. 14
Hollow Camshaft Technology

ASSEMBLED CAMSHAFTS :

Another, comparatively recent, camshaft weight reduction technology is the "Assembled Camshaft" which consists of a hollow steel tube with either forged or sintered steel cam lobes assembled onto it together with pulley mounting flanges, timing sensor rings, etc. Typical weight saving is 30-35% compared with a solid cast iron camshaft.

Although the process is inherently more expensive than a cast iron camshaft, and doesn't offer any real advantages for sliding contact applications, it allows the use of steel camlobes to take the high contact stresses of rolling contact applications at lower cost than a solid steel shaft, which might otherwise have been required.

Assembled camshafts are best suited to designs with bearing caps where the bearing journals can be machined directly onto the pre-hardened tube, as opposed to "tunnel" designs which would require the bearing journals to be assembled onto the tube, as well as the cam lobes.

There are probably as many patented assembly techniques as there are manufacturers of assembled camshafts. Two examples of techniques in current production, used by Krupp-Presta and Suko (MWP), are shown in Figs 15 & 16. Another common technique is to expand the tube inside the cam lobes by pushing or pulling a ball or mandrel through it or by internal hydraulic pressure.

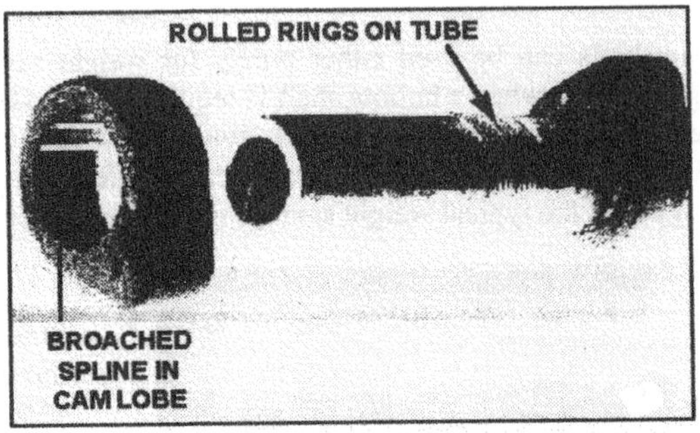

Fig. 15

- Krupp-Presta Assembly Technique :
    - Broached spline in cam lobe pressed over rolled rings on tube.
    - Tube is induction hardened locally before assembly, and finish ground after assembly.
- Cam Materials:
    - Hot-forged steel.
    - Sintered steel.

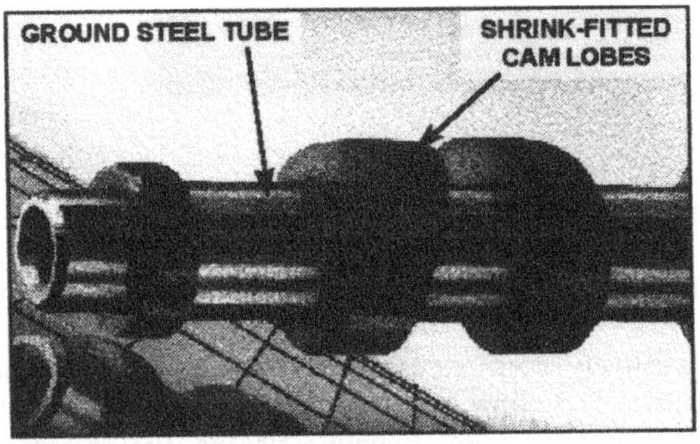

Fig. 16

- Suko Assembly Technique :
    - Cam lobe shrink-fit onto ground steel tube.
    - Tube hardened and ground to finished bearing size before assembly.
- Cam Materials:
    - Hot-forged steel.

- Sintered steel.
- Net-shape sintered steel - doesn't require finish grinding and allows use of small re-entrant radius forms.

A reduction in the cost penalty of an assembled camshaft over a cast iron shaft, in a chain or gear drive engine, can be achieved by fixing the sprocket or gear directly to the steel tube using the same fixing technique as the cam lobes. An example of this, used in the Ford Duratec 2.5L V6 gasoline engine cam is shown in Fig 17. This technique is more suited to gasoline than diesel engines because the cam timing is less accurately controlled than with a separate sprocket and the build up of tolerances can lead to valve to piston clash in a diesel engine which has very close valve to piston proximity for combustion and emission control.

Fig. 17
Ford Duratec V6 Camshafts (Krupp Presta)

LIGHTWEIGHT VALVE TRAIN COMPONENTS :

Engine designers are looking more and more to lightweight tappet and valve train components to reduce the effective reciprocating mass of the valve train. This can allow lighter valve spring loads to be used, which in turn can give reduced friction and improved economy and emissions. Reduced valve train mass can also allow higher cam accelerations to be used for performance or efficiency improvements.

This means that materials and technologies which were once the prerogative of the racing engine designer are now being explored for less exotic, higher volume, applications.

Fig. 18

Fig 18 shows two applications for lightweight materials in a direct-acting bucket tappet valve train :

➤ An aluminium alloy tappet, as used in the V8 Jaguar engine, weighs about 40% of a steel tappet. When used in an aluminium cyl head, because the expansion coefficients are closely matched, running clearances can be optimised over a wider range of temperatures. The tappet skirt is normally iron plated for use in an aluminium head.

➤ Ceramic (normally silicon nitride) tappet adjusting shims also weigh about 40% of steel shims. Initial test results indicate that silicon nitride will accept almost twice the contact stress of a 100Cr6 steel shim running against same chilled cast iron cam lobe material. Ceramic shims must be polished (to <0.1Ra) to prevent abrasive wear of the cam lobe. Further polishing significantly reduces friction, to levels comparable with low-friction coatings applied to steel shims. The main obstacle to the use of ceramic shims, to date, has been their cost and most of the current work on ceramics is aimed at achieving cost-effective production processes.

➤ Three other weight-saving opportunities in the valve train are :
   ➤ Conical (tapered) valve springs. These have a reduced diameter at the top end which allows a smaller, lighter, valve spring retainer to be used whilst maintaining a large enough diameter at the bottom end to fit over the valve stem seal.
   ➤ Reduced valve stem diameters. Whereas several years ago Ø8mm was the norm for passenger car engines with Ø7mm just starting to appear, Ø6mm is now the norm for direct-acting bucket tappet engines with Ø5mm starting to appear in production and even smaller diameters being developed.
   ➤ Hollow valve stems. These are generally used in conjunction with sodium cooling for highly rated engines where increased heat transfer away from the valve head / seat area is required. Their prohibitive cost makes them very much a last resort for volume production engines.

OPERATION OF 2 VALVES BY 1 CAM LOBE :

The operation of 2 valves with one cam lobe has the potential to reduce valve train costs, and overall engine weight, by halving the number of camlobes and followers used. Figs 19 shows a design for a direct-acting arrangement. This uses a bridge piece, containing 2 hydraulic lash adjusters and one roller, which slides on 2 guide pins. Oil for the hydraulic lash adjusters is fed through one of the guide pins and the bridge piece.

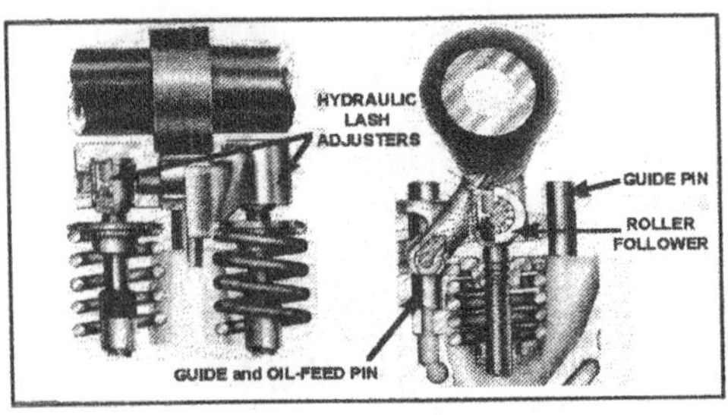

Fig. 19
INA Concept Design - Bridge Piece with Roller and 2 HLA's

Fig. 20
Forked Roler Rocker with Hydraulic Lash Adjusters
(Chrysler Neon 2.0 Litre Gasoline Engine)

Fig 20 shows a forked rocker arm design, again with 2 hydraulic lash adjusters and a single roller.

The cost benefits of halving the number of cam lobes is partially offset by the fact that the force at the cam-to-follower interface is doubled, because each cam is carrying the force of 2 valve springs, and a wider cam lobe and follower will probably be required.

4 Valves per Cylinder Operated by a Single Camshaft :

The ultimate engine designer's goal is to operate all the valves in the engine with one camshaft. Whilst this is relatively easy in gasoline engines with inclined valves, it is more difficult to achieve in a 4-valve per cylinder diesel engine where the valves have to be vertical for optimum combustion performance. One example, for a large, push-rod, diesel engine is shown in Fig 2. Fig 21 shows an example for a small (2.0L) diesel engine, using standard hydraulic bucket tappets, operated in pairs by a chilled cast-iron "slipper".

The advantages of this design are its simplicity and low cost.

The disadvantages are the inherently high friction associated with sliding contact, hydraulic lash adjuster, valve trains and the careful attention to lubrication and cam profile design which is required to avoid wear problems.

Fig 22 shows a concept for achieving the same result using a bridge piece sliding on a pin, with a roller follower and 2 mini hydraulic lash adjusters.

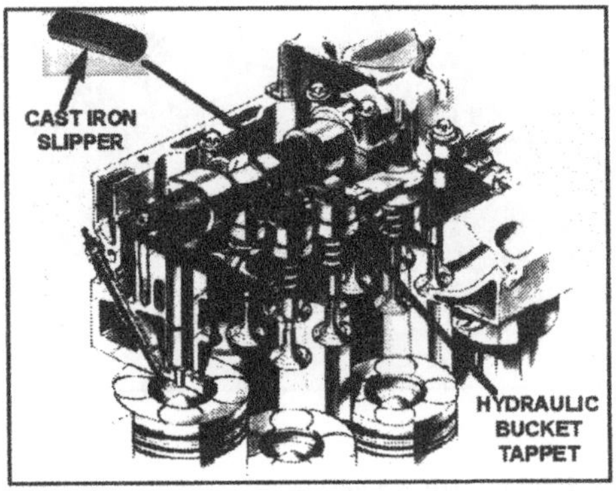

Fig. 21
4 Valves per Cylinder operated by Single Camshaft
(GM Ecotec 2.0l DI Diesel Engine)

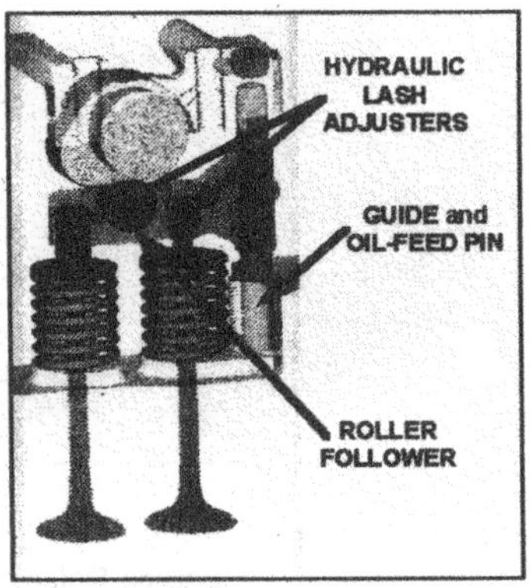

Fig. 22
INA Concept Design - Bridge Piece with Roller and 2 HLA's

SUMMARY

- Increasing customer and legislative demands for increased fuel economy, without performance sacrifice, and reduced emissions have promoted a shift away from sliding contact, hydraulic lash adjuster, valve trains using heavy, solid, cast-iron components.
- Modern valve train technology is moving towards reduced friction, with mechanical adjustment for bucket tappets or roller followers with hydraulic lash adjustment.
- Total engine weight reduction can be achieved by the use of hollow cast-iron or assembled steel camshafts.
- Alloy and ceramic valve actuation components reduce reciprocating, as well as total engine, weight. Reduced reciprocating weight allows "lighter" valve springs to be used which also give an associated friction reduction.

ACKNOWLEDGMENTS

I would like to thank the following companies for their assistance in providing material for this article :

- INA Bearings.
- Krupp Presta.
- MWP.
- G. Clancey.

Ring zone lubrication: implications for reduced emissions and enhanced engine life

M FOX
Department of Chemistry, DeMontfort University, UK

ABSTRACT

The i/c engine lubricant in the piston ring zone operates under severe conditions of temperature, pressure, shear and oxidation. The condition of the lubricant in the piston ring zone is substantially degraded relative to that of the lubricant in the sump.

Trends in lubricant requirements are driven by both commercial and environmental improvement forces. Two important effects are shown to arise from current trends in ring pack design:

- The mass of the cylinder bore oil film has been substantially reduced whilst the mass of lubricant within the ring pack has been maintained or increased,

- The residence time of ring pack lubricant has increased by a factor of approx. ten.

Less lubricant will be present for longer in the piston ring pack.

The performance of ring packs in for rpm and bmep is complex needs to be considered in models.

INVESTIGATING LUBRICATION IN THE PISTON RING ZONE

The mission of the Lubricant Research Centre is to investigate lubrication of the ring zone of operating i/c engines. This zone is a crucial interface which has to provide enough lubricant to minimise wear but prevent excess lubricant transport into the combustion chamber. At the same time, the lubricant must withstand severe thermal and shear degradation yet complete a pressure seal. An underlying mission of the Lubricant Research Centre is to develop an integrated science and technology approach, recognised as 'complex machines and materials lubricated by complex fluids'. Our major facilities are:

Petrol	**Low Speed, Single Cylinder**	**High Speed, Multi-Cylinder**
	Petter W-1's	Mercedes M-102, 111
Diesel	**Single Cylinder**	**Multi-Cylinder**
	Petter AA-1's, AVB	CAT 3406B VW1431

Our other research interests include base oil oxidation, the chemical and physical properties of lubricant samples and development of improved analytical methods.

Our approach to investigating ring zone lubrication are based upon the physical models shown in Fig 1, where (a) sets out the basic piston/oil film/bore temperature and heat flow concepts and (b) develops this further for a piston and ring pack. Fresh, colder, oil is transported up to the ring pack from the lower bore wall and hot oil is displaced back down to the sump. Figure 1(b) also shows the four sampling points used on the CAT engine piston, numbered 1-4.

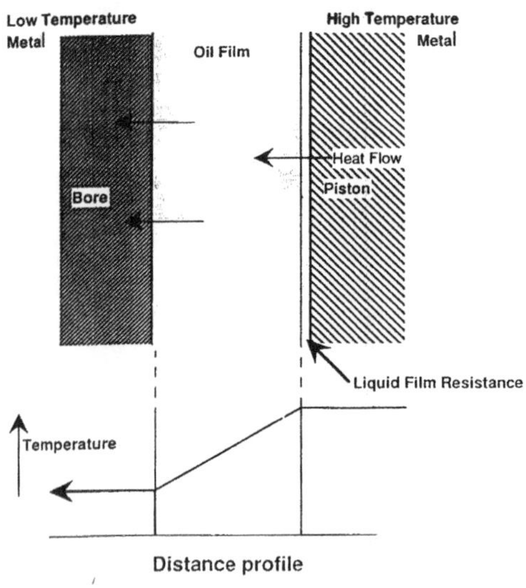

Fig 1 (a)

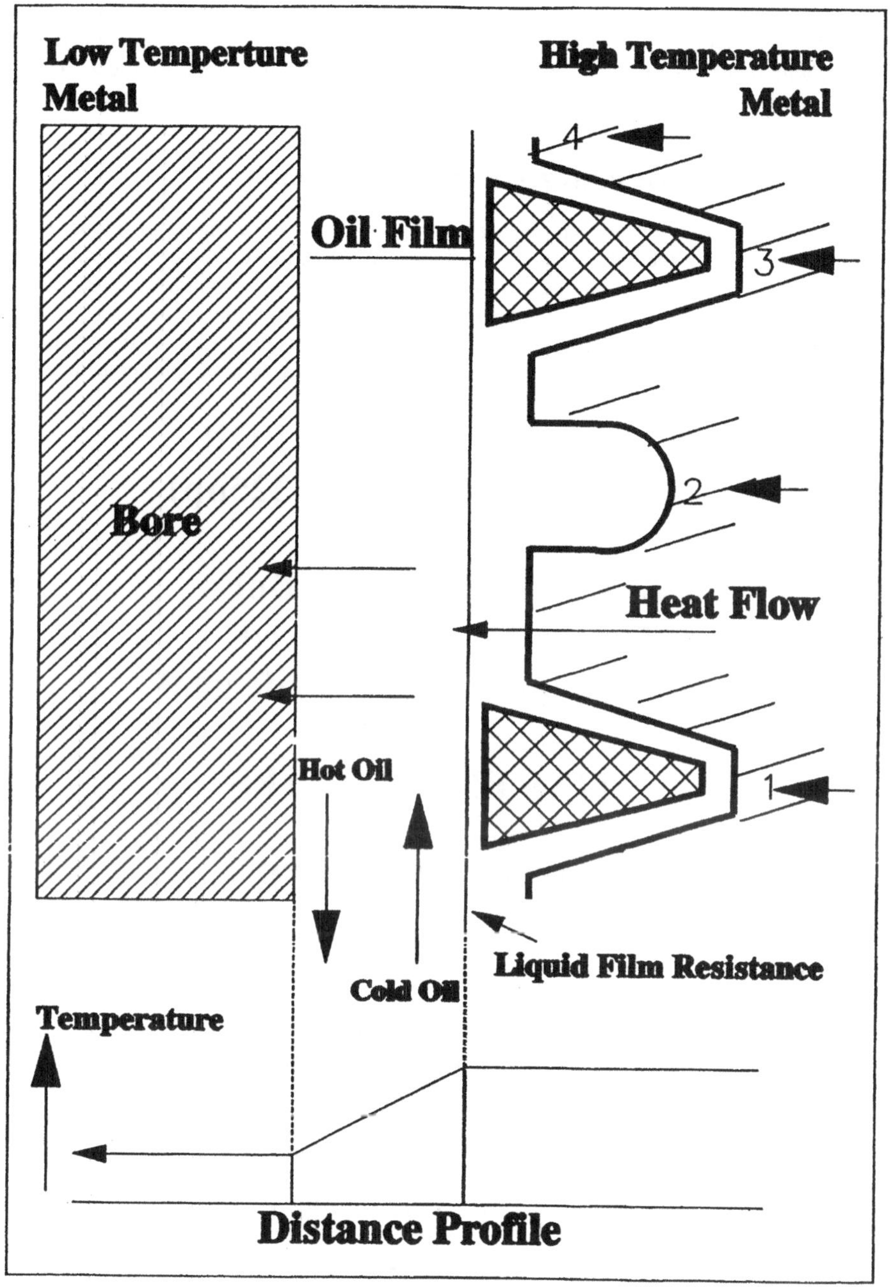

Fig 1 (b) Schematic diagram of piston sampling positions and oil flow

The work described is for the CAT-3406B engine, which was part of (US)DoE/NASA sponsored 'LE55 High Efficiency Engine' project, aiming to give 55% thermodynamic efficiency with reduced emissions and longer service life goals. This translates into a 40 tonne payload hauled at 88km/hr consuming 12mpg. Subsequent work is reported separately in terms of physical and chemical effects.

PHYSICAL ASPECTS OF RING ZONE LUBRICANT SAMPLES

The ring zone sampling system has been applied to two sets of sampling positions, either positions 1 and 3 (from the back of the second and first ring grooves, respectively) and from positions 2 and 4 (from the accumulator groove and the base of the crown land, respectively). The number of sampling positions has been limited to two in any one experiment so as to minimise the amount of lubricant sampled. The system conducts the sample to the exterior of engine as a high velocity aerosol. The physical aspects of the ring zone samples considered are set out in Fig 2.

PHYSICAL EFFECTS FOR
CAT 3406B (15Ltr, 6cyl, 325hp)

LE55 PROJECT, US-DoE
55% efficiency, reduced emissions, longer service life goal

- GAS FLOW RATES

- OIL FLOW RATES

- RING PACK CHANGES

- TRANSFER RATES, SUMP TO RING PACK

EP PROPERTIES OF THE LUBRICANT

Fig 2

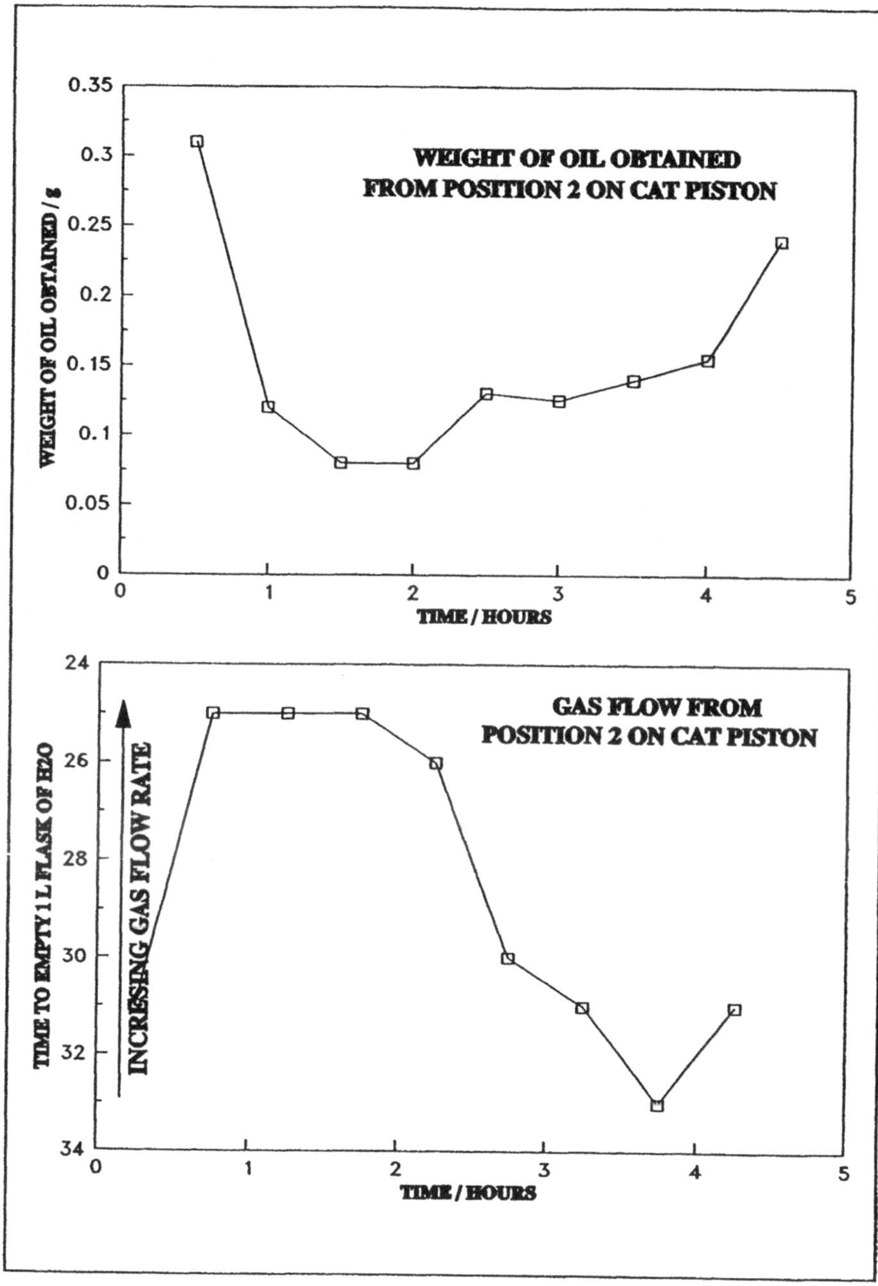

Fig 3 Variations in oil mass and gas flow with CAT 3406B engine run time

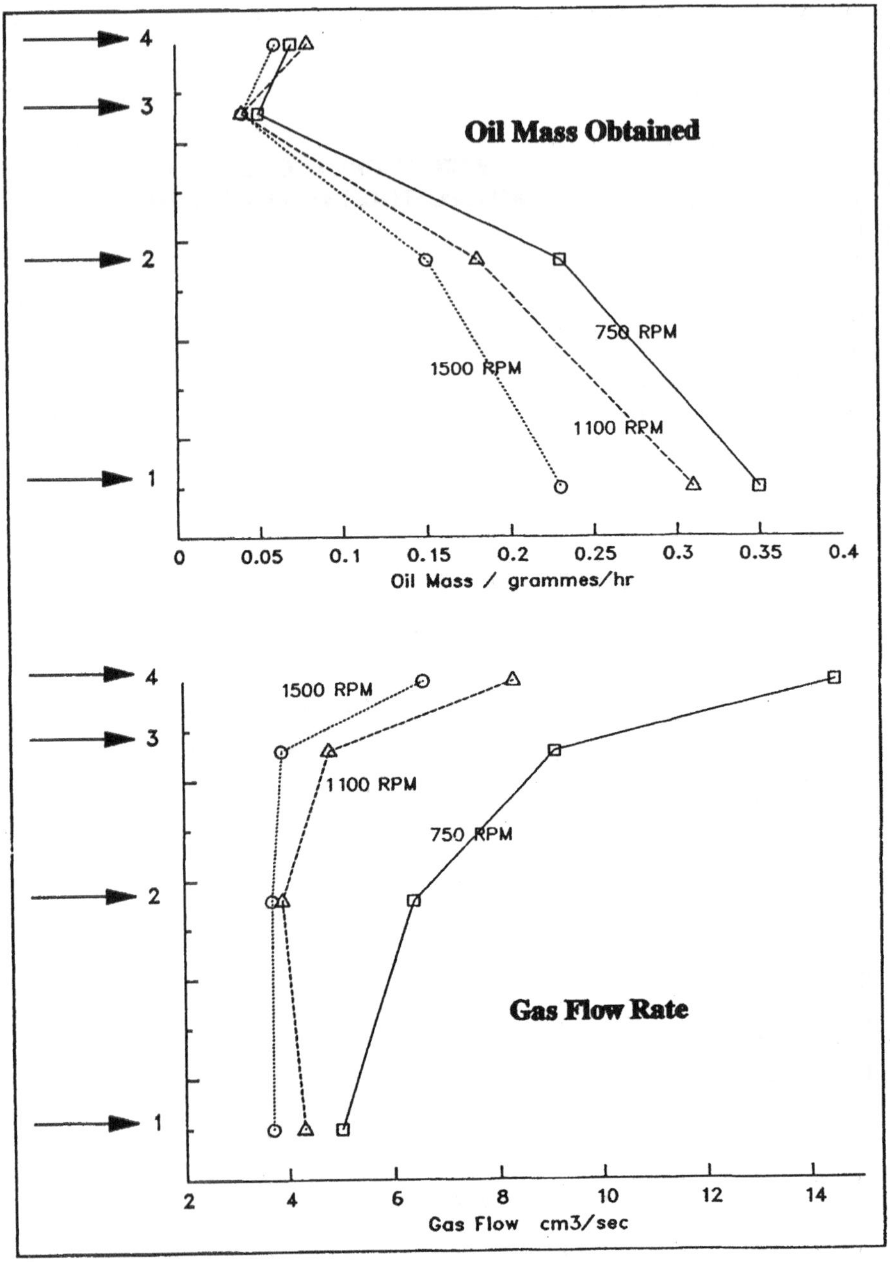

Fig 4 Oil mass and gas flow as a function of sampling position and engine RPM for the CAT 3406B

Lubricant droplets are separated from the gas by impaction in a glass vial. The lubricant sampling rate varies for several reasons such as the oil viscosity, engine load (bmep), rpm and a cyclic time variation, Fig 3. The reasons for this variation are not clear; contenders might be ring rotation, with gaps coming into coincidence or thermal cycling of the piston and ring assembly. All of the subsequent work reported here has been obtained at the same point of the cyclic variation.

As a first investigation, Fig 4 shows a decrease in the oil mass and gas flow rate from successive sampling positions up the piston towards the crown, except for a relative increase at position 4, the base of the crown land, compared to the first ring groove, position 3. Increasing the rpm and load gives a decrease in the oil mass flow rate from the first ring.

A detailed investigation of the oil mass and gas flow rates sampled from the first and second ring grooves, Figs 5 and 6, shows unexpected behaviour in initially declining with rpm and then increasing to a plateau. Oil flow from the second ring groove increases to a plateau. For both ring grooves, the dominant effect on oil flow rate appears to be rpm rather than bmep, except at the higher bmep values. The 3-D plot for the first ring flow rate is 'noisy' because of the errors involved in sampling very small volumes. The gas flow rates vary in a similar manner to that of the oil flow. Both plots show that oil mass and gas flows through the ring pack vary in a non-linear manner with rpm and load. It will be interesting to see how ring pack design can alter these performance characteristics. Further, if detailed models of ring pack performance can encompass or predict these variations.

Oil Mass Data

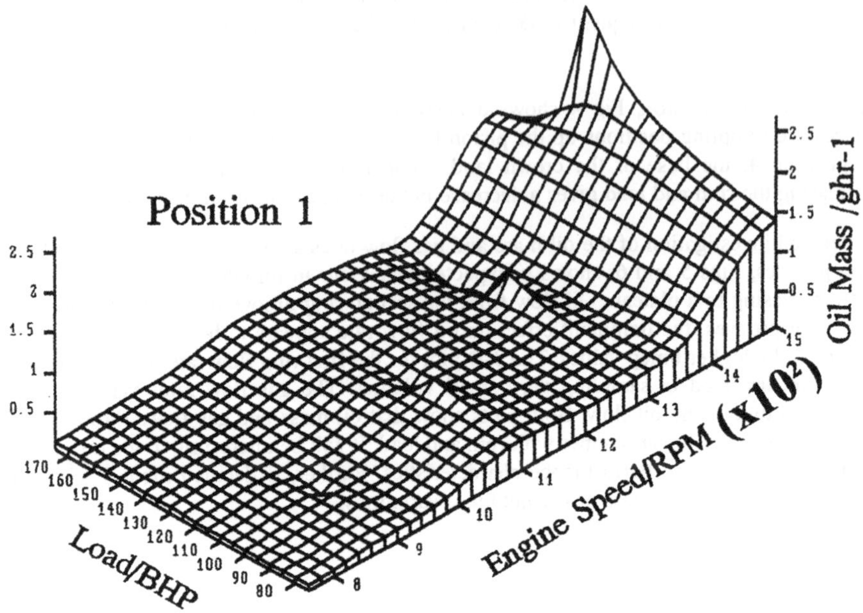

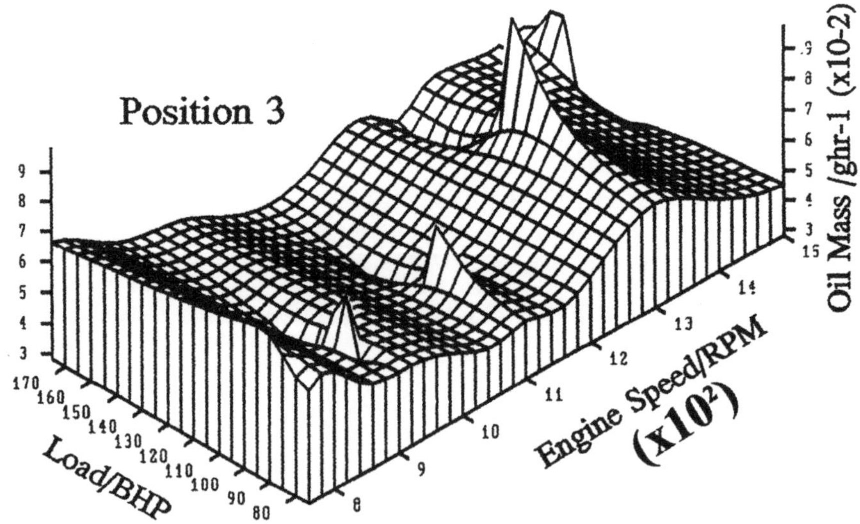

Fig 5

Gas Flow Data

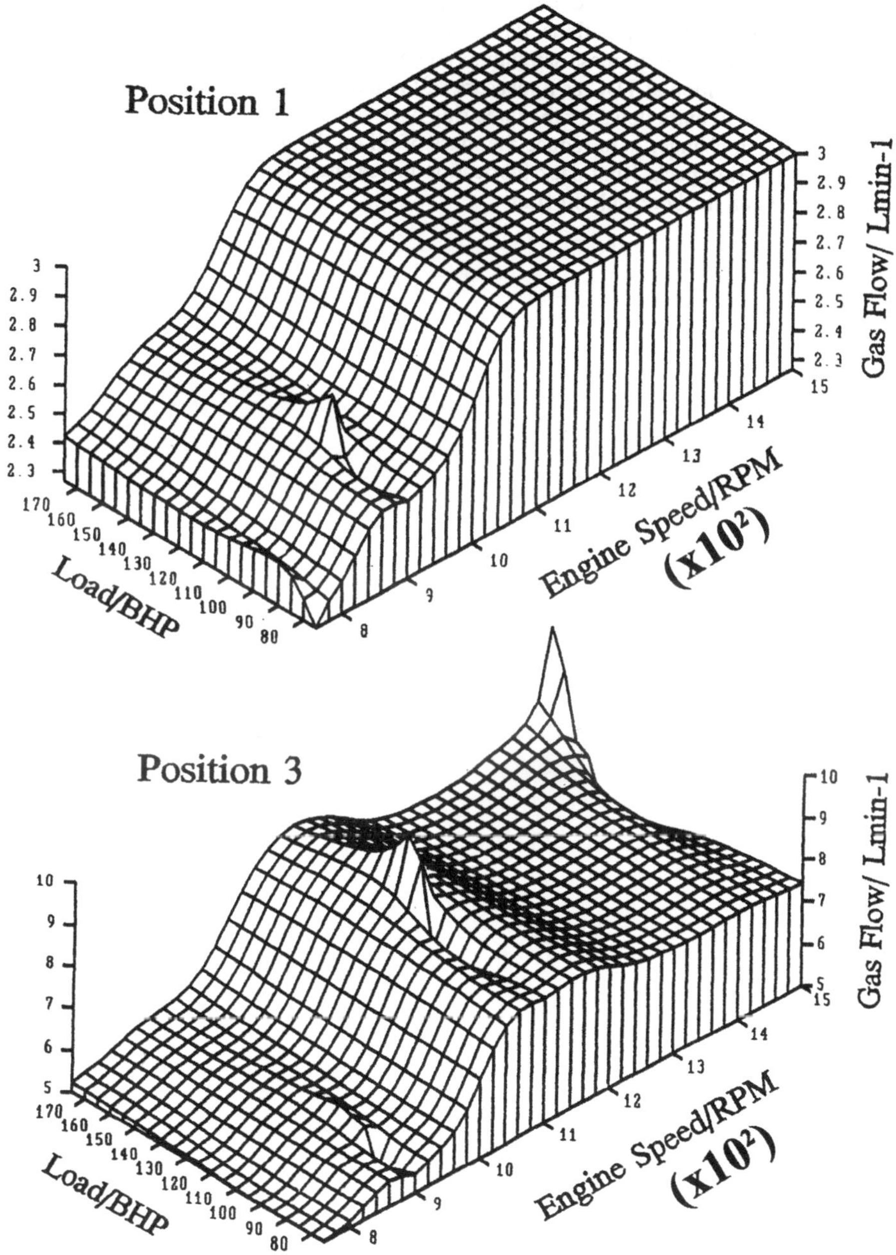

Fig 6

Ring pack design changes have been driven by the need to reduce emissions arising from ingress of lubricant to the combustion chamber. Three successive sets of piston rings for the CAT 3406B have been tested under identical conditions. The ring sets correspond to designs immediately prior to the (US)1991 diesel emission control regulations, post-1991 and post the (US) 1994 emission control regulations. Each set has detailed dimensional changes to alter its performance. An additional oil mass sampling system used in this work was positioned at the first ring reversal position at TDC as a measure of the mass of the oil film on the bore wall. Comparison of the oil mass flows for the three ring pack sets, Figs 5, 7 & 8, shows that the net effect of the design changes is to reduce the mass of oil on the bore wall but to increase the mass of oil in the ring pack. Ring pack lubrication increasingly relies upon the oil contained within it and less on the bore wall oil film.

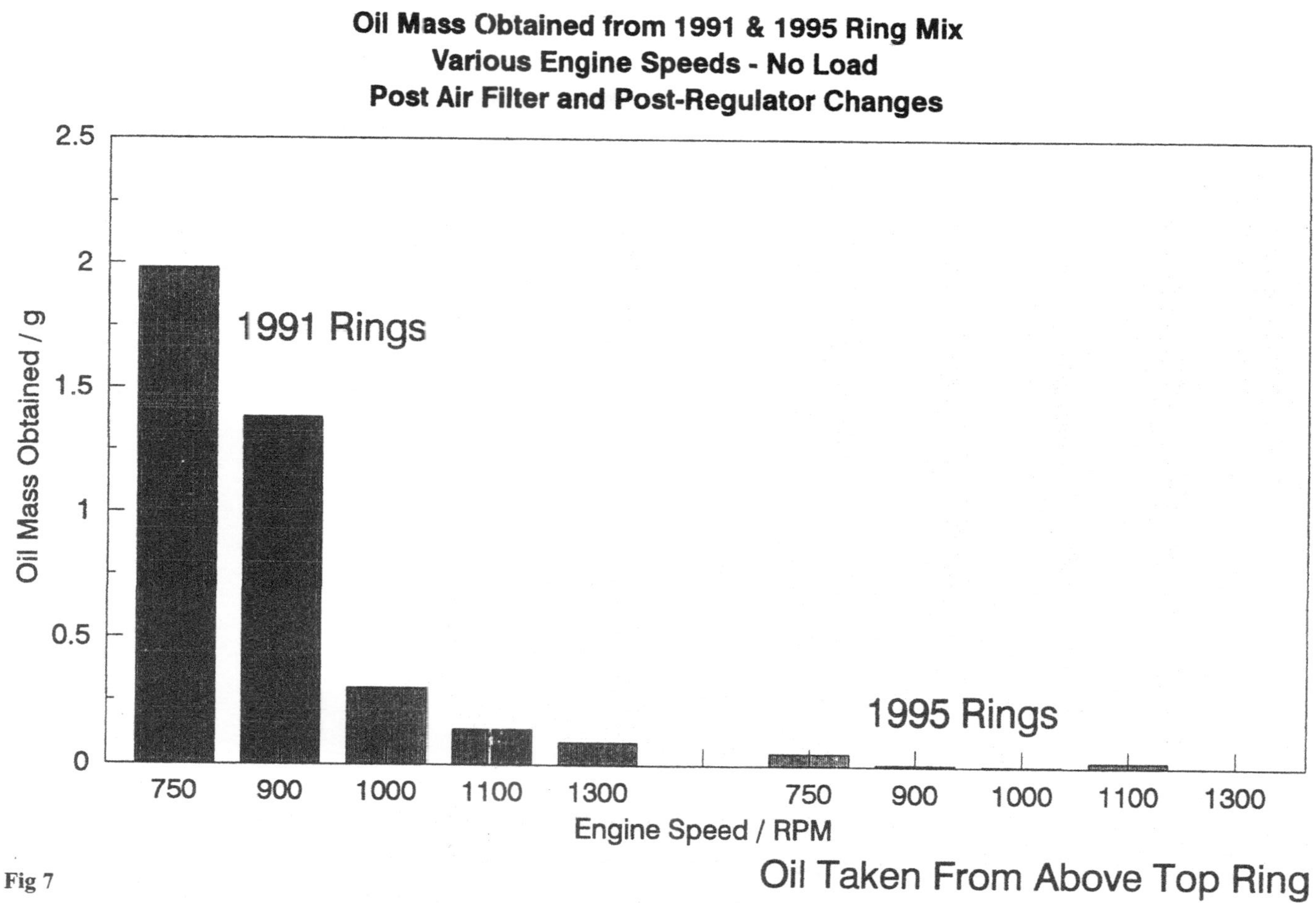
Oil Mass Obtained from 1991 & 1995 Ring Mix
Various Engine Speeds - No Load
Post Air Filter and Post-Regulator Changes
Oil Mass Obtained / g
2.5
2
1.5
1
0.5
0
1991 Rings
1995 Rings
750
900
1000
1100
1300
750
900
1000
1100
1300
Engine Speed / RPM
Oil Taken From Above Top Ring

Fig 7

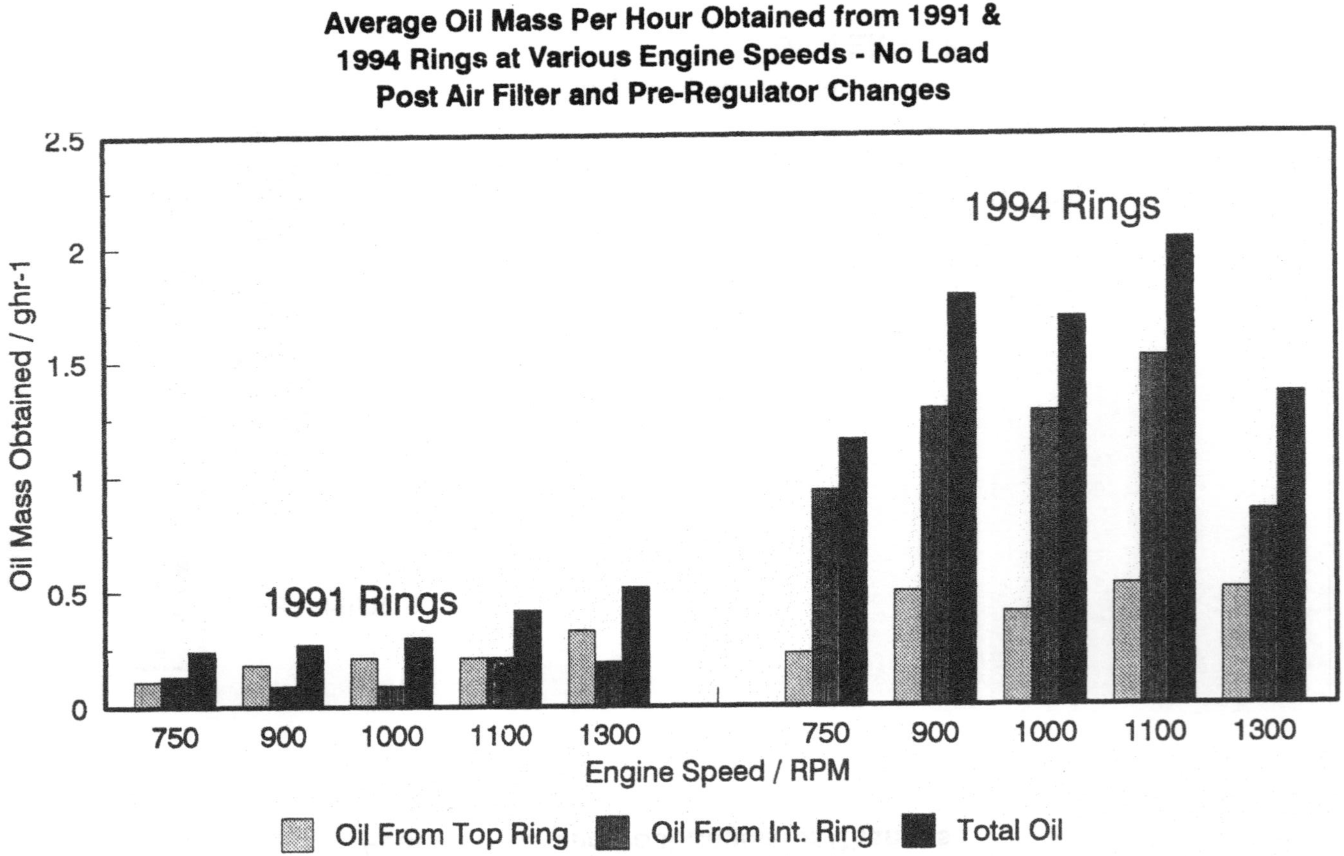
Average Oil Mass Per Hour Obtained from 1991 &
1994 Rings at Various Engine Speeds - No Load
Post Air Filter and Pre-Regulator Changes
Oil Mass Obtained / ghr-1
0
0.5
1
1.5
2
2.5
1991 Rings
1994 Rings
750
900
1000
1100
1300
750
900
1000
1100
1300
Engine Speed / RPM
Oil From Top Ring
Oil From Int. Ring
Total Oil

Fig 8

An important concept is the 'residence time' of an oil particle in the ring pack. This can be defined in several ways but one which is readily amenable to measurement is the time taken for a 'marker' to travel from an equilibrated sump to establish an equilibrium concentration in the piston ring zone. We have measured this previously for engines with relatively 'old' ring designs, namely an Petter AA-1 (0.22ltr) and a CAT 1Y-73 (2.2ltr), both giving equilibration times of about 3-4 minutes. When repeated for the CAT 3406B, with the pre-1991 ring pack, a magnesium sulphonate marker gave a residence time to the first ring groove of 26 minutes, eight times that of the value previously measured for the older engines. The corresponding times to the second ring groove are six minutes, 16 minutes to the accumulator groove and 72 minutes for sampling at the base of the crown land. There is a linear relationship for residence times at sampling positions 1-3, but not for position four, Fig 9. As a result of ring pack design changes driven by reductions in pollutant emission, the major implication is that less lubricant is present in the piston ring zone and that it remains there for a substantially increased time.

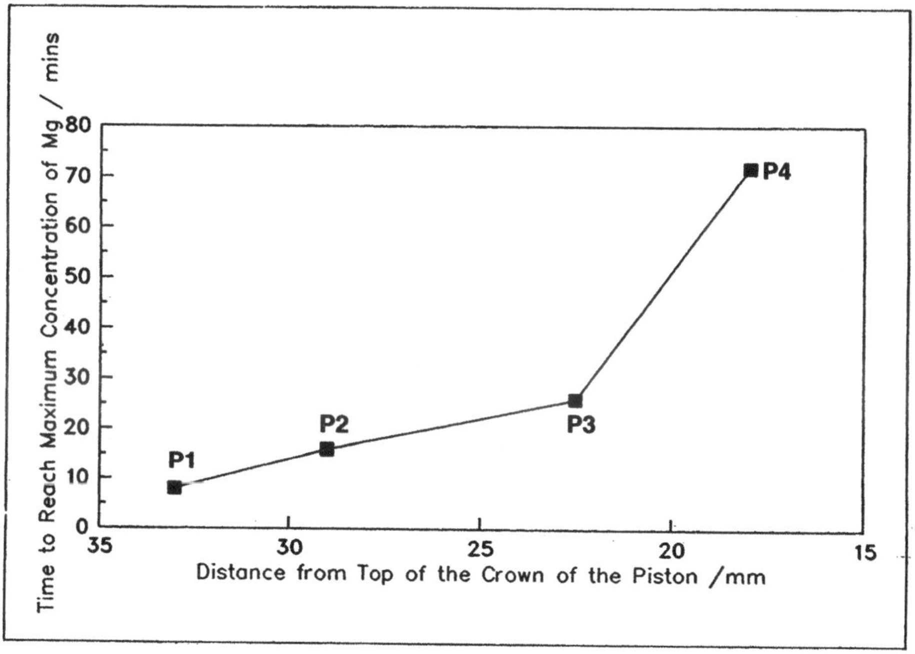

Fig 9 Comparison of sampling position to time to reach maximum concentration of magnesium transfer from the sump of the CAT 3406B, commercial 15W40 lubricant

One method of assessing the extreme pressure (ep) properties of ring zone lubricant samples is the 'Four-Ball Tester' device. Fig 10 shows how the initial increase in scar diameter declines to lower loads for both the mono- and multi-grade lubricant samples obtained from the ring zone, relative to the results for the fresh lubricants. The weld points decline similarly. These results demonstrate that the ep properties of ring zone lubricant samples are significantly degraded relative to their original, fresh, condition.

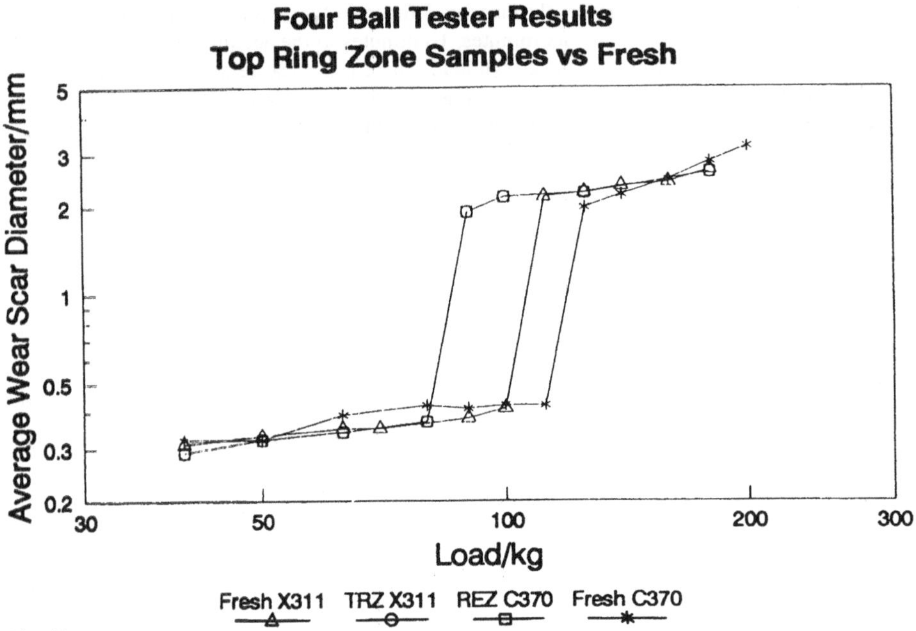

Fig 10

There is a continuing and substantial increase in the particulate levels of sump lubricants. Samples obtained from the four different sampling positions shows several particulate size distributions, broadly centered on ~1, 6 and 30m, Fig 11. There appears to a series of equilibria governing the inter-conversion growth between these distributions, this effect is now being investigated separately. It is also clear that whilst dispersants work very well in the sump of engines, dispersants are made ineffective and destroyed by the severe physical and thermal conditions in the piston ring zone. This has considerable implications for maintaining piston cleanliness, reducing ring sticking and increased engine service life.

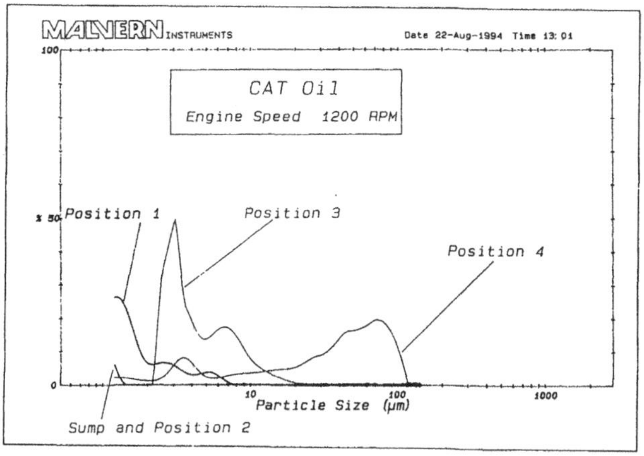

Fig 11 Comparison of particle size with respect to sampling position on the CAT 3406B piston, commercial 15W40 lubricant

CHEMICAL ASPECTS OF RING ZONE LUBRICANT SAMPLES

Formulated lubricants for i/c engines are complex fluids, approx. 80-90% base oil and 20-10% of a mixed additive pack. Analyses of piston ring zone lubricant samples presents difficulties because the samples are so small, typically less than 1gm/hour, varying upon the sampling position and lubricant formulation. A typical analytical scheme, Fig 12, shows the relative oil sample masses used for each component analyis. The chemical effects observed are set out in Fig 13.

CHEMICAL EFFECTS

ANALYSES

INFRARED (FT-IR)

MICRO-VISCOMETRY

THERMO-GRAVIMETRIC ANALYSIS (TGA) FOR RELATIVE VOLATILITY

BASE & ACID NUMBERS

Fig 12

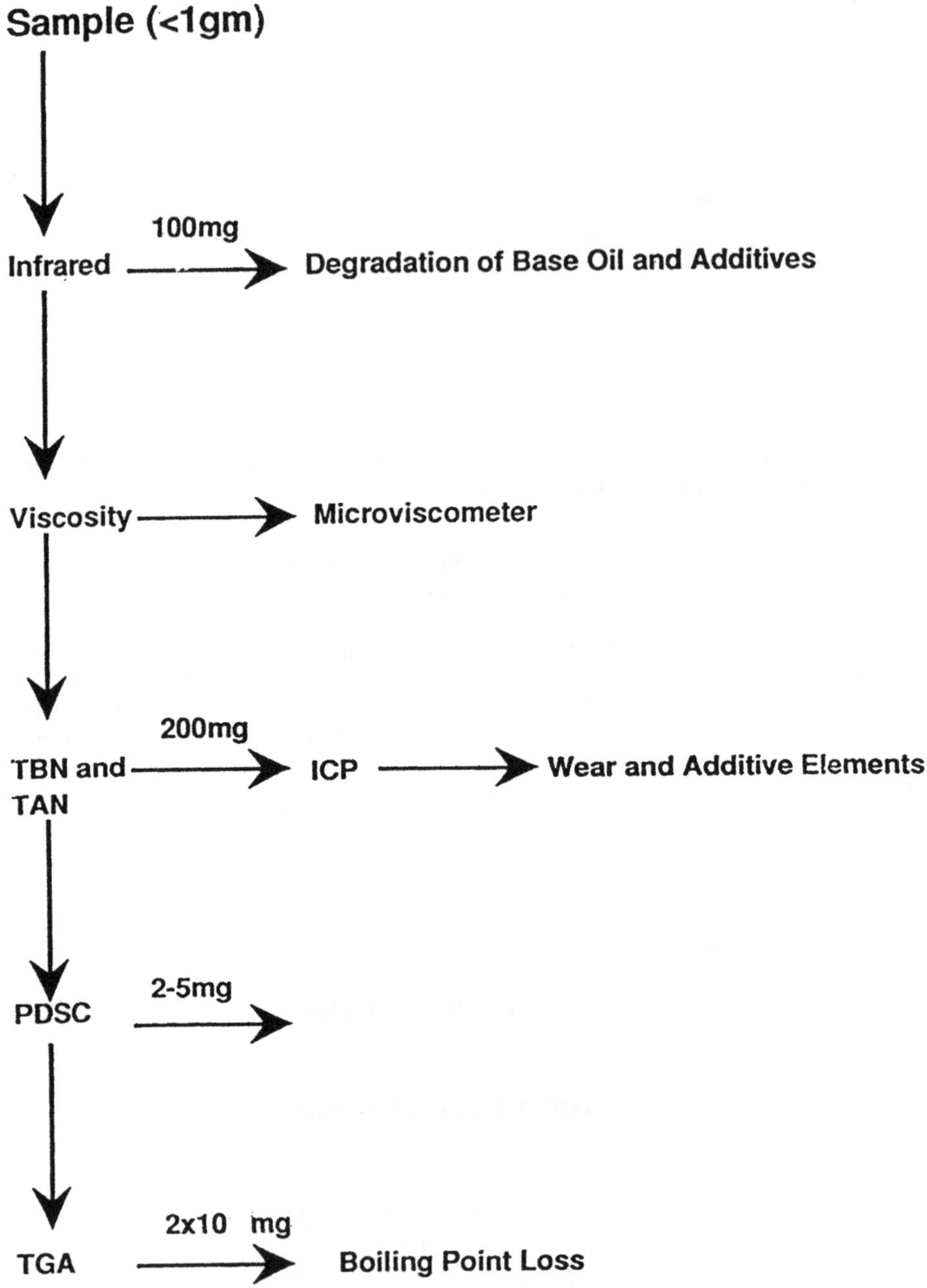
Sample (<1gm)
100mg
Infrared
Degradation of Base Oil and Additives
Viscosity
Microviscometer
200mg
TBN and TAN
ICP
Wear and Additive Elements
2-5mg
PDSC
2x10 mg
TGA
Boiling Point Loss

Fig 13

Infrared spectra of fresh lubricant samples are complex and become more so with service life degradation, particularly for piston ring zone samples. Successful samples in an engine test run show a progressive series of changes, Fig 14, but there is not much that can be obtained quantitatively from these spectra. If the spectrum of the original, fresh, lubricant are subtracted from each of these successive test run spectra, normalised spectra are obtained, Fig 15.

Defined regions can be integrated to give changes in such chemical functions as carbonyl, >C=O, mainly indicative of hydrocarbon oxidation. Similarly, the degradation of the anti-oxidant present can be followed as the lubricant is distributed through the ring pack, Fig 13. The composite plots within Fig 16 show that the anti-oxidant components measured as P=S and P-O-C absorption bands of the ZDDP additive degrade in the lower part of the ring pack whilst protecting the hydrocarbon base oil. The onset of carbonyl oxidation in the upper part of the ring pack occurs once the anti-oxidant has been used up.

Similarly, depletion of the anti-acid carbonate additive occurs throughout the ring pack, Fig 17. Sulphate/C-O-C formation increases, reflecting the neutralisation of sulphur oxides (from fuel sulphur) in the combustion gases and nitration also occurs increasingly up through the ring pack.

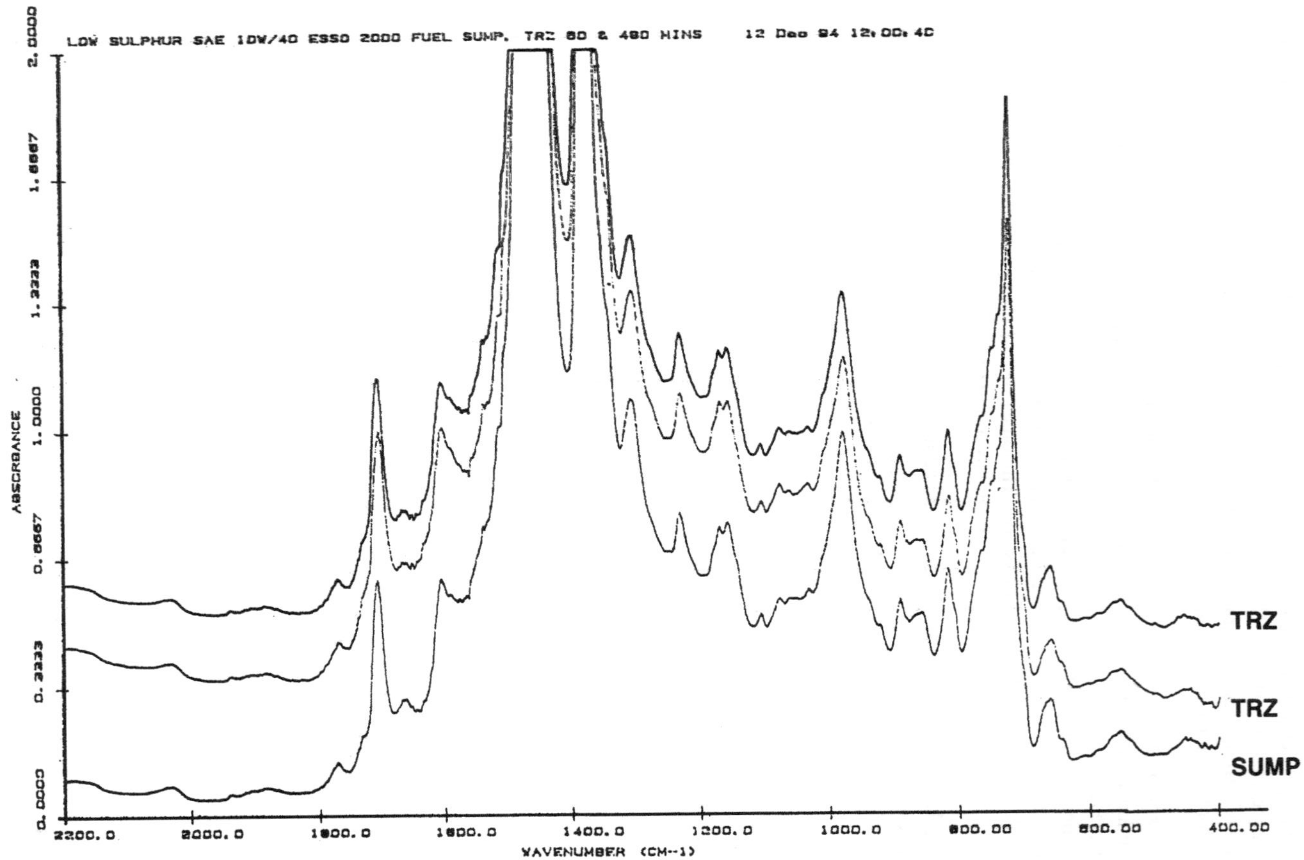

Fig 14 FTIR spectrum of low sulphur 10W40 lubricant TRZ samples, 0.18% fuel, @ 200°C, for 1hr and 8hr TRZ and sump

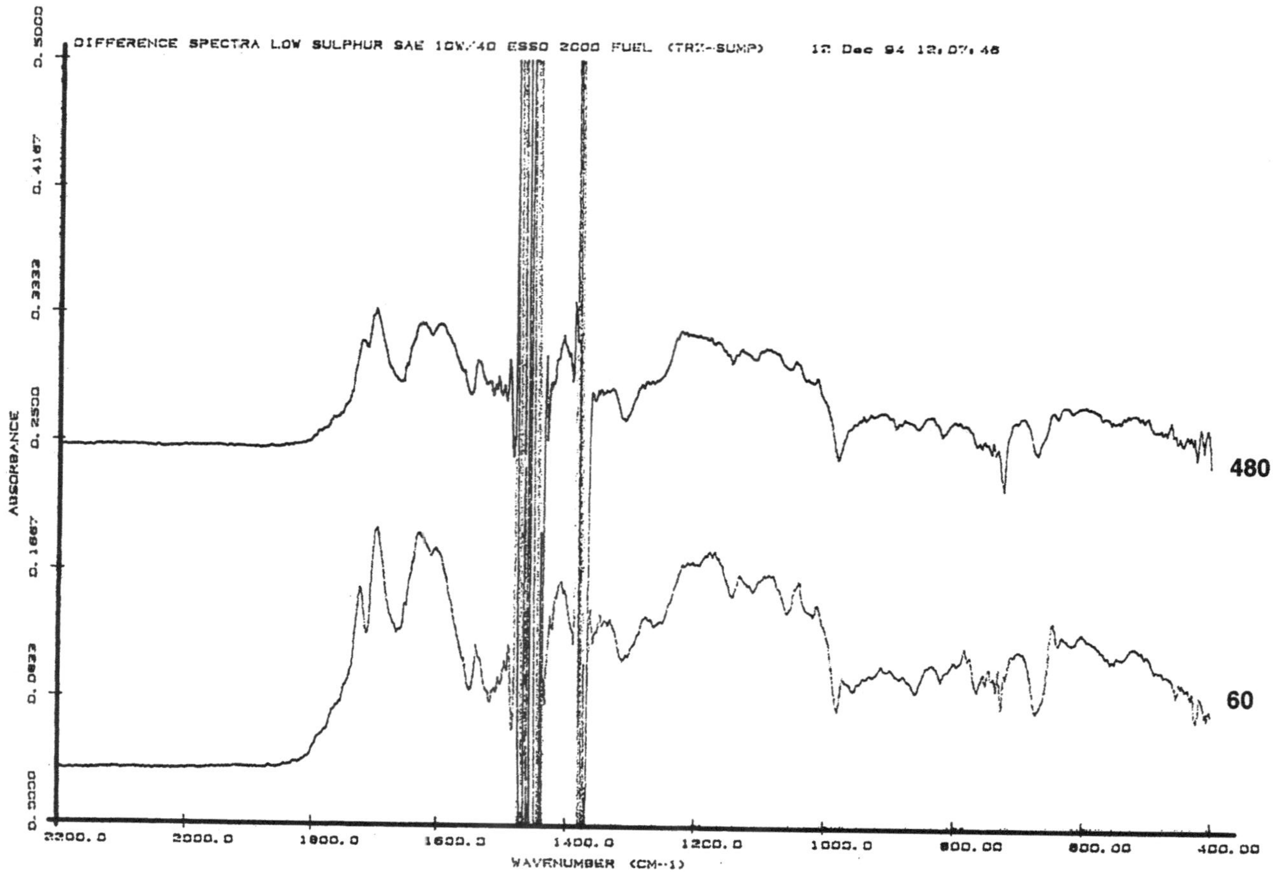

Fig 15 FTIR difference spectra of low sulphur 10W40 lubricant TRZ samples, 0.18% fuel, @ 200oC, for 1hr and 8hr TRZ

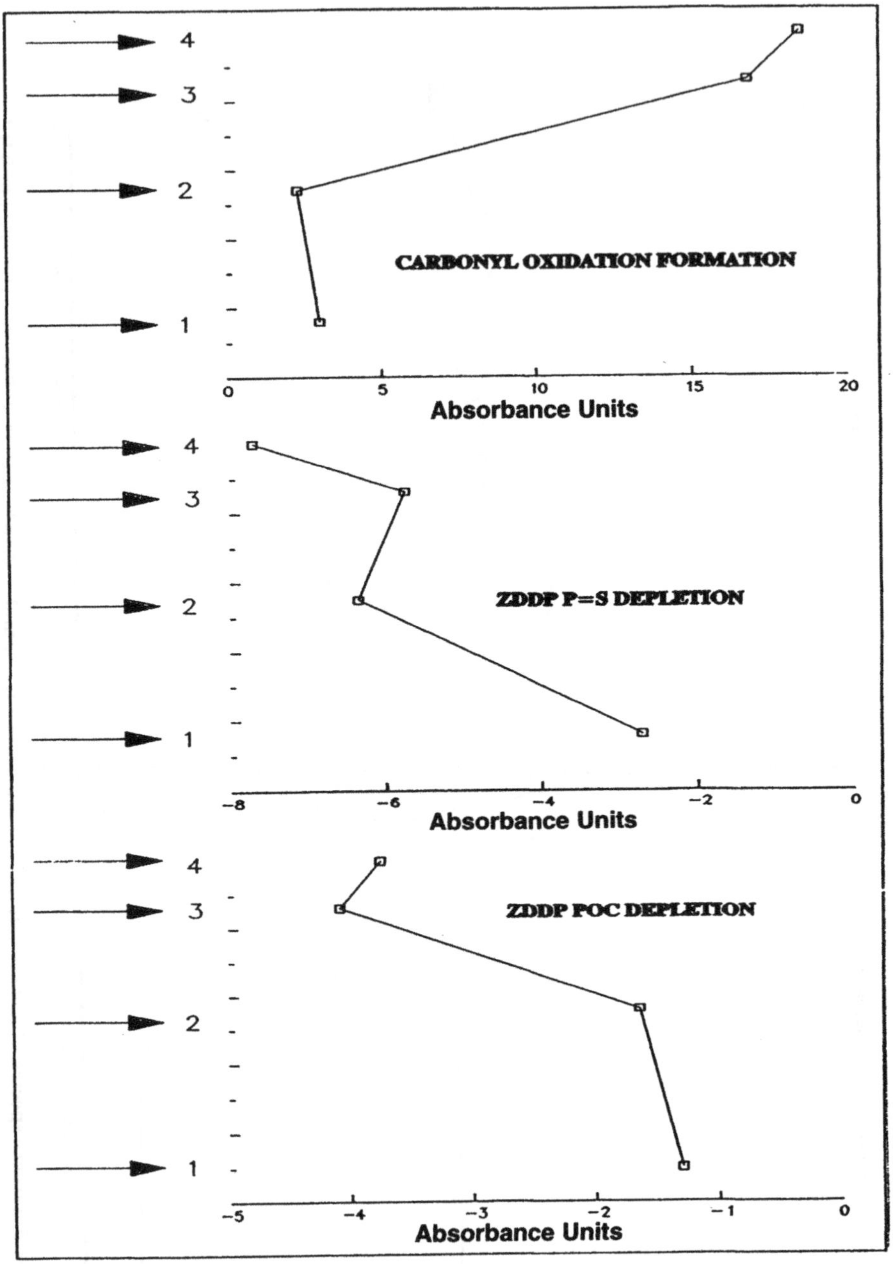

Fig 16 Comparison of FT-IR integral peaks for commercial 15W40 lubricant sampled at 4 positions on the CAT 3406B piston

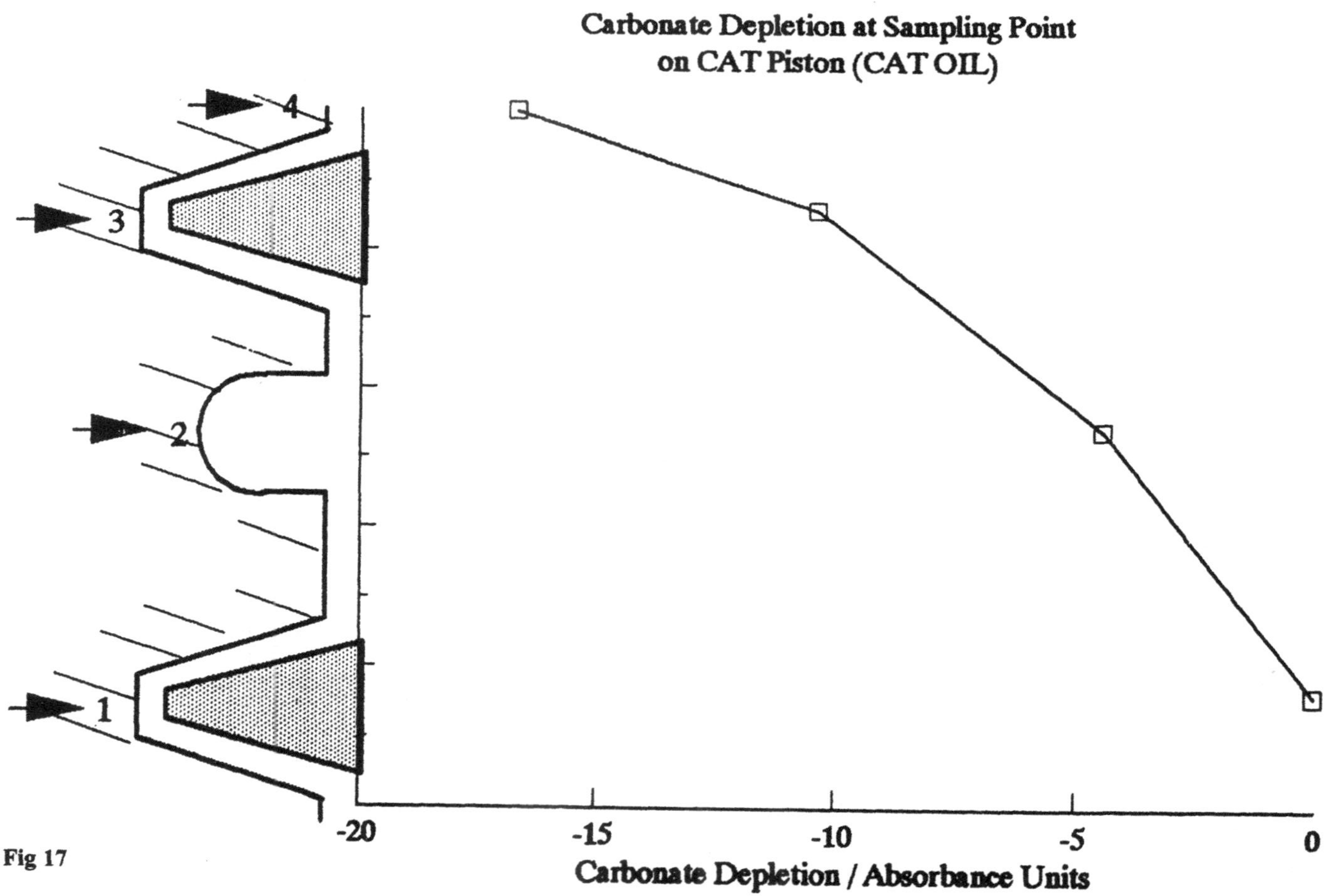
Carbonate Depletion at Sampling Point
on CAT Piston (CAT OIL)
4
3
2
1
-20
-15
-10
-5
0
Carbonate Depletion / Absorbance Units

Fig 17

Two interesting and complementary plots are the Viscosity Index of the lubricant sampled at different points in the ring pack and their relative volatility, the latter determined by a micro-ThermoGravimetric Analysis, or TGA. Whilst initial inspection of Figs.18 & 19 indicate some confusion, more detailed analyses show that they are indeed complementary. Moreover, they show that the relative volatility and viscosity indices of the samples obtained from behind the first and second ring are substantially different from the samples from the surface of the piston, from the base of the crown land and from the accumulator groove. These results demonstrate that the condition of the lubricant is different in the ring grooves from the lubricant film on the bore wall, as collected on the surface of the piston.

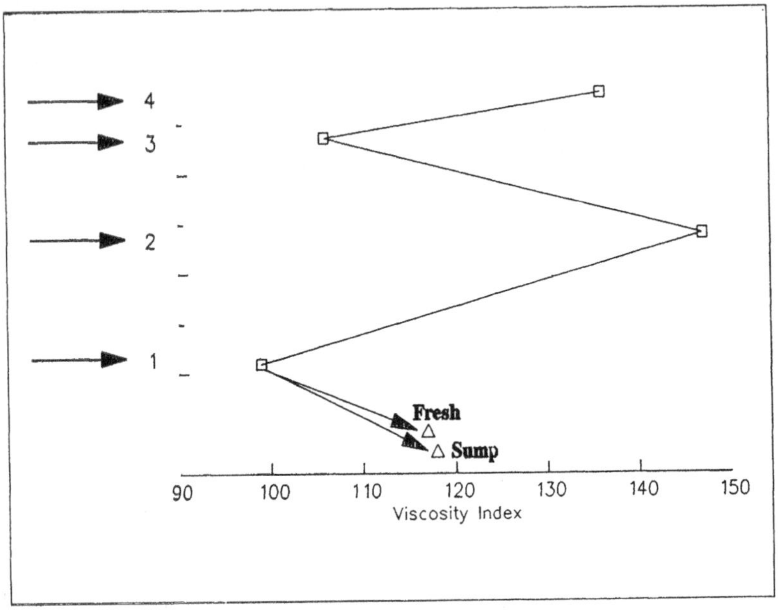

Fig 18 Comparison of viscosity index with respect to sampling position on the CAT 3406B piston, commercial 15W40 lubricant

SUMMARY

The work outlined above serves to demonstrate that the lubricant in the piston ring zone operates under severe conditions, as summarised in Fig 20. The condition of the lubricant in the piston ring zone is substantially degraded relative to the lubricant in the sump.

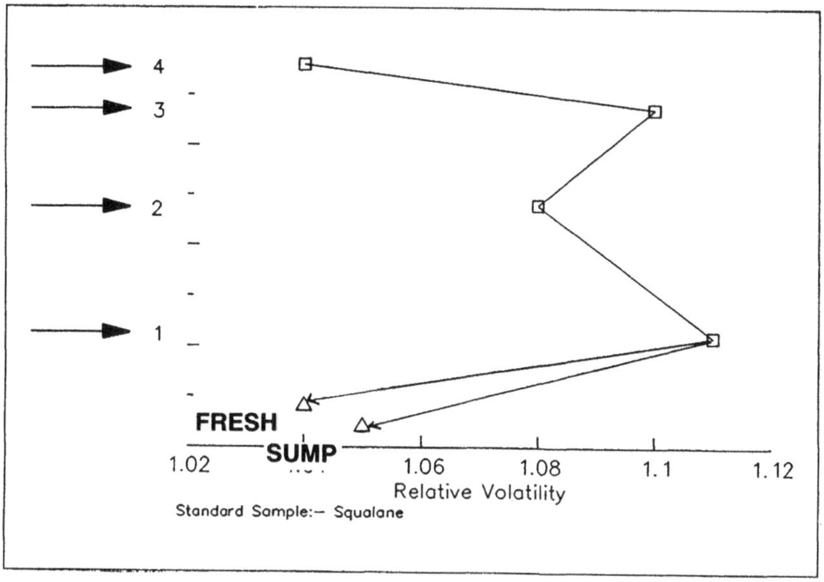

Fig 19 Comparison of TGA results with respect to sampling position on the CAT 3406B piston, commercial 15W40 lubricant

Further, trends in lubricant requirements are driven by both commercial and environmental improvement forces. Two important effects are evident arising from current trends in ring pack design:

- that the mass of the oil film on the cylinder bore has been substantially reduced whilst the mass of lubricant within the ring pack has been maintained or increased,
- that the residence time of lubricant in the ring pack has been increased, close to a factor of ten.

Therefore, less lubricant will be present for longer in the piston ring pack.

The dynamic behaviour of ring packs in respect of rpm and bmep show complex behaviour which must be addressed in modelling studies.

PISTON RING ZONE

SEVERE CONDITIONS

- HIGH TEMPERATURE

- VARYING PRESSURE

- HIGH SHEAR RATES

PHYSICAL DEGRADATION OF LUBRICANT

CHEMICAL DEGRADATION OF LUBRICANT

Fig 20

MODEL

Automotive lubricants: today's choice, tomorrow's possibilities

A J MOORE
BP Oil Technology Centre, UK

SUMMARY

In addition to minimising friction and wear, engine oils must prevent the build-up of deposits on engine surfaces and help control the pollutants emitted as exhaust. As service intervals lengthen, oil-sumps become smaller and operating temperatures rise, lubricant design must continually change to provide the performance needed by modern engines. Some of the directions in which lubricant design must develop in the near future are discussed in the light of the conflicting demands imposed by key lubricant specifications.

BACKGROUND

The last few years have seen a considerable tightening in the ways lubricant formulations are defined, tested and approved. Under the rules which now govern lubricant development, the composition of a formulation must be defined at the start of an expensive approval process and then retained with little deviation until the process is either successfully completed or abandoned. Combined with the increasingly rapid rate at which lubricant specifications are changing, the high cost of lubricant development is doing little to encourage longer-range thinking about the needs of future engine designs. Activity is becoming increasingly short term in focus and directed toward the fine balance needed to satisfy the often conflicting demands of engine cleanliness, durability and fuel efficiency.

The rapid changes that are taking place in lubricant specifications reflect, most notably, the ever-increasing competitive and legislative pressures on OEMs to improve fuel efficiency and reduce exhaust emissions (see Fig. 1). Whilst such considerations were of relatively little interest to lubricants marketers until quite recently, they have now become the principal determinants of the way in which formulation design is approached. Governmental targets imposed on OEMs in the USA and voluntarily agreed in Europe powerfully influence the way in which lubricant specifications are evolving. These, in turn, determine the way in which lubricant suppliers must develop their product lines.

Lubricant quality approval systems

For major oil companies operating on a global basis, two lubricant approval systems are of particular importance. Each demands that lubricants meet a set of physical and chemical criteria and also fulfil certain performance requirements in a suite of engine and laboratory tests. In the USA, the basic approval system is that operated by the American Petroleum Institute, the API. Oils meeting the current range of API requirements may claim to be of API SH quality, the succeeding quality designation to be introduced shortly being API SJ. Whilst oils claiming the API quality designation are not obliged to meet a fuel efficiency (FE) requirement, they may claim to be of Energy Conserving Level I (ECI) or Energy Conserving Level 2 (EC2) status if the meet the appropriate performance standard in the ASTM Sequence VI bench engine test. Closely related to the API quality standard is the international standard, ILSAC GF-2, relevant to products intended for the American and Japanese markets. ILSAC GF-2 requirements are identical to those of API SH but with the addition of a mandatory fuel efficiency requirement based on the ASTM Sequence VI-A engine test.

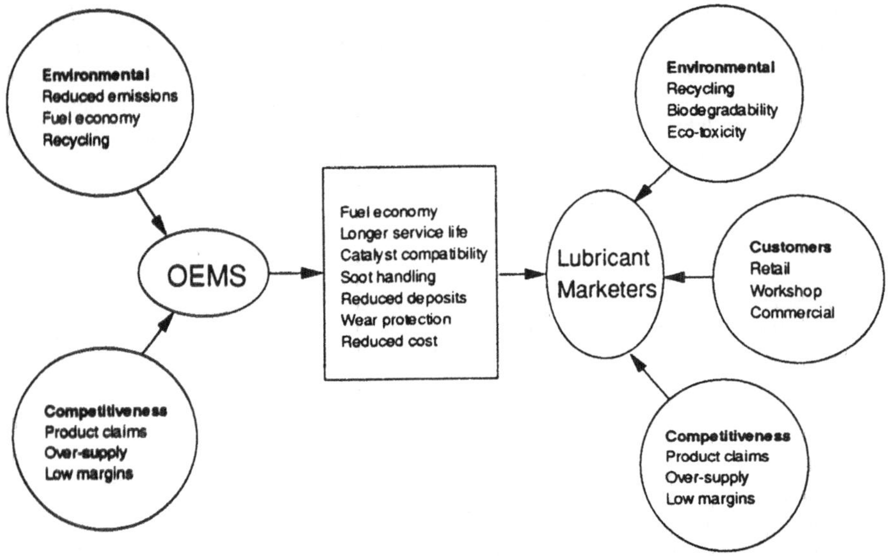

Figure 1. Business pressures faced by original equipment manufactures and lubricant suppliers

The CCMC quality standard which held sway in Europe until about two years ago consisted essentially of the prevailing API standard combined with and a number of additional engine and laboratory tests demanded by European OEMs. A new set of quality standards since introduced by ACEA, the industry body which now represents automotive manufacturers in Europe, overlaps to a greatly reduced degree with API/ILSAC requirements. Only two engine tests, the Sequence IIIE and Sequence VE, now remain in common. The divergence between the two quality standards has increased the technical complexity of lubricant development and greatly increased development costs. Mergers now taking place between both oil companies and additive manufacturers reflect the increasingly harsh economic climate that has developed.

Market influences

An important factor behind the difference between European and American lubricant requirements is the high proportion of passenger car diesel engines in the European market. Since the workshops that service passenger cars prefer to use a single lubricant for gasoline and diesel engines, European lubricants must be able to satisfy diesel requirements. In consequence, the high temperature/high shear rate (HTHS) viscosity demanded by ACEA for the two top tier classes; of lubricant remains fixed at the relatively high value of 3.5 mPas. The ILSAC quality standard, in contrast, is driving HTHS viscosities down towards 2.6 mPas for SAE xW-20 grades of lubricant and 2.9 mPas for SAE 5W-xx grades. Even lower values will be embraced by US and Japanese OEMs if engine design can be suitably modified.

FUEL EFFICIENCY PERFORMANCE

The need for improved fuel efficiency

Prompted originally by national economic and energy security concerns in the USA, the drive towards further improvement in automotive fuel efficiency is now propelled by concerns about atmospheric pollution and greenhouse gas emissions. Cars and trucks currently account for 25% of CO_2 emissions in the USA and are expected to represent an even larger proportion in years ahead. To enforce FE improvements by automotive manufacturers, the US government introduced Corporate Average Fuel Efficiency (CAFE) targets in the late 1970s and supported them with severe financial penalties for companies which failed to meet their targets. The threat was effective. Fig. 2 illustrates how General Motors rose to the challenge of CAFE requirements over the period 1979 to 1994. Average fuel economy increased from about 18.6 mpg to 27 mpg over this fifteen year period, a rise of 45%.

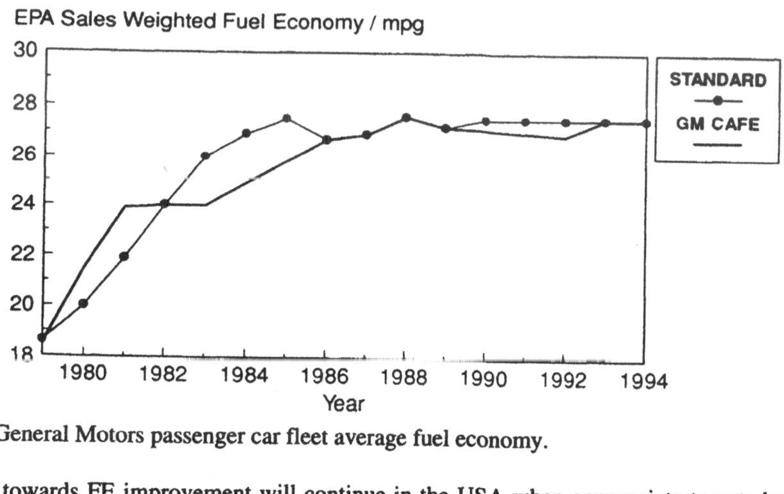

Figure 2. General Motors passenger car fleet average fuel economy.

The drive towards FE improvement will continue in the USA when appropriate targets have been agreed. These are likely to bear a close resemblance to the "technically feasible" bands, in Fig. 3, proposed by the National Academy of Sciences. The 20% rise from 27 mpg to about 34 mpg over fifteen years reflects the increasing difficulty of the task

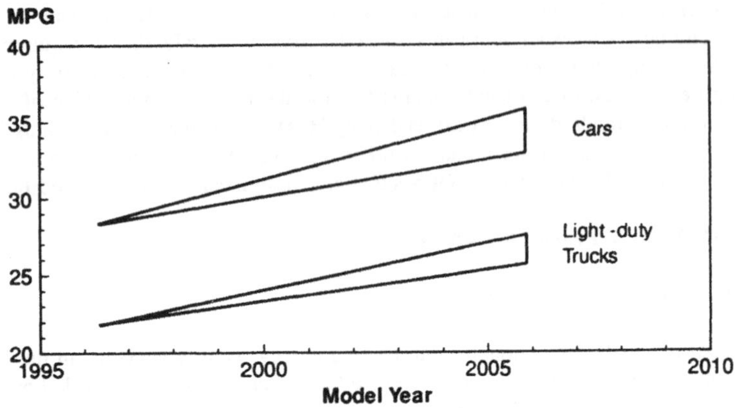

Figure 3. CAFE expectations: National Academy of Sciences "Technically Achievable" bands

Lubricant influences on fuel efficiency

Lubricant-related friction losses in automotive engines are largely centred on the piston assembly, main bearings and valve train. Although operating conditions vary widely in these areas, it nevertheless proves possible to construct relatively simple empirical relationships between lubricant properties and FE performance. A good representation of lubricant influences is typically given by an expression of the form

$$FE = a - b.\mu - c.\eta - d.\alpha \qquad (1)$$

where a, b, c and d are constants, μ a boundary friction coefficient, η the HTHS viscosity and α a composite measure of pressure-viscosity response and the additional shear thinning that occurs at shear rates above 10^6 s^{-1}.

An example of how equation (1) can rationalise the behaviour of lubricants in FE tests is given in Fig. 4 where predicted and determined values of fuel efficiency improvement are compared for the Sequence VI-A engine test. This approach to the analysis of lubricant behaviour provides a means by which lubricant designers may investigate and understand the effects of alternative formulation approaches.

Boundary friction properties are dictated by additive chemistry. Friction control is most cost-effectively exercised through the use of friction modifiers, additive components having the specific role of reducing boundary friction. Fig. 5 shows that a conventional organic friction modifier will reduce the boundary friction coefficient from about 0.13 to 0.11, an improvement that will result in a modest but nevertheless useful gain in FE performance. Unfortunately, it also tends to induce a deterioration in anti-wear performance, a potentially unacceptable handicap in the European market. In the USA it represents less of a problem.

More marked friction reductions can be obtained using molybdenum dithiocarbamate (MODTC) additives. These materials are multi-functional in nature and both reinforce anti-wear

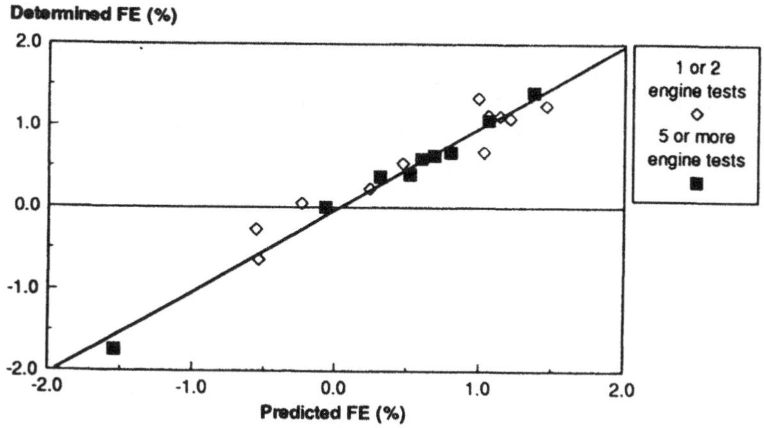

Figure 4. Comparison between predicted and determined FE performance in the ASTM Sequence VI-A engine test.

characteristics and improve antioxidancy. As Fig. 5 demonstrates, best results tend to be obtained when MODTC additives are used in combination with primary alkyl ZDDPs rather than the secondary alkyl ZDDPs most commonly used in Europe. An attraction of a move from secondary to primary ZDDP usage in European formulations is the improvement likely to be found in passenger car diesel engine performance.

A downside of the multi-functional nature of MODTC additives is that they tend to be depleted relatively rapidly. As the oil ages in service, FE performance consequently deteriorates. Further work is needed to determine how this tendency can be contained.

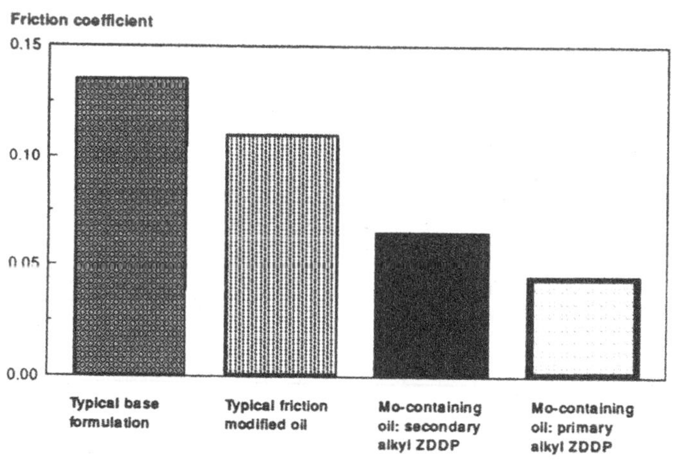

Figure 5. Representative boundary friction coefficients for automotive engine oils

The effect of viscous properties on power loss is, from the formulators point of view, quite complex. It has been known for many years that FE performance is strongly influenced by high shear viscosity and the parameter η, in equation (1), remains the single most important lubricant property. As will be seen, its magnitude depends, firstly, on the viscosity of the base oil blend, secondly, on the thickening effect of the dispersant inhibitor (DI) additive package and, thirdly, on the temporary viscosity loss characteristics associated with the polymeric VI improver. The relationship η to other important viscous characteristics is illustrated in Fig. 6. In general, viscosity varies with shear rate between two limits. The upper limit is the low shear viscosity of the full formulation and the lower limit the viscosity of the DI solution. η, the measure of HTHS viscosity demanded by the SAE J300 viscosity standard is the value determined at 150°C and the shear rate of $10^6 s^{-1}$. It usually falls below the low shear viscosity but well above the limiting high shear viscosity defined by the properties of the DI solution.

Since shear rates in modem engines can rise well above the value of $10^6 s^{-1}$ at which η is determined, this value of viscosity may overestimate the true effective value. Equation (1) provides some compensation for this deficiency through the influence of the third parameter α. Oils having the same pressure-viscosity coefficient will differ in a values according to their shear thinning characteristics. A high polymer content will cause a to be relatively low and FE to hence be relatively high. Although, with this prospect in mind, it is helpful if the viscosity of the DI system is held to as low a value as possible by the additive manufacturer, the current trend is for the viscosity of additive packages to rise. This is problem is not an insoluble one but may require the design of dispersant additives to be re-visited.

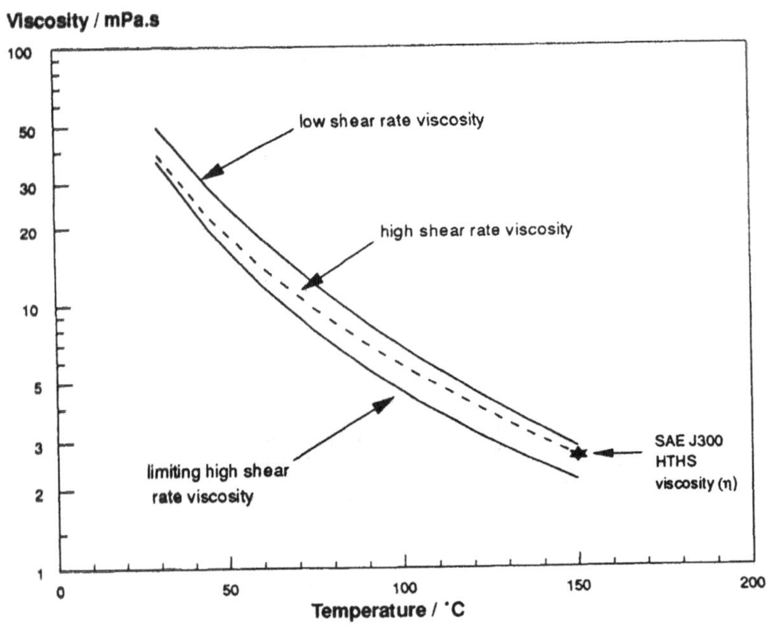

Figure 6. Viscous characteristics of polymer-containing engine oils

EXHAUST EMISSIONS

Lubricants affect exhaust emissions both through their influence on fuel efficiency and their potentially deleterious effect on catalytic converters. Phosphorus is one of a number of elements known to poison catalyst materials and is restricted by ILSAC GF-2 to a maximum concentration of 0.1% w/w. As emissions targets become more demanding in the future and better performance is needed from catalytic converters, further restrictions on phosphorus content may follow. Under ILSAC GF-3, a phosphorus limit as low as 0.06% w/w may be imposed. Alternatively, a direct test of catalyst performance may be introduced if an acceptable test method can be devised.

The Jaguar AJ/V8 engine

T N CRISP
Jaguar Cars Limited, UK

INTRODUCTION

A totally new engine is a tremendously important and costly event for Jaguar, and this is only the fourth new engine in our history.

The first real Jaguar engine code-named XK, engine was conceived during the last World War and went into production during 1948 in the XK120. This was a 3.4 litre 6 cylinder engine with twin overhead camshafts and hemispherical combustion chambers which produced 160 bhp in a car capable of 120 mph. The cylinder head design concept of the cams operating directly on bucket tappets was unique in a production engine at that time, but is now the preferred choice for high performance engines and remains a feature of the new AJV8. Of course it was the XK engine which went on to power 5 Le Man wins in Jaguar XK'C', and XK'D' types when it delivered well over 300 bhp. in less powerful guise it remained in production until 1986.

The next engine milestone was the 5.3 litre V12 engine introduced in 1971 which set new standards in terms of refinement and performance, and in 6 litre form is still in production today. This engine too won racing honours in winning the European Touring Car Championships in an XJS in 1984

During the 1970's the oil crisis caused us to consider the design of lighter and more economical engines for the following decade. The AJ6 engine introduced in 1983 was one of the first 4 valve, 6 cylinder all aluminium engines. It had a capacity of 3.6 litres with 4 valves per cylinder, as well as a 2.9 litre 2 valve derivative. This was approximately 100 lbs lighter than the XK engine and developed 225 bhp. In 1990 the stroke was increased to give a capacity of 4 litres and in its current product form develops a power output of 245 bhp and this engine together with a shorter stroke 3.2 litre version continues to power our XJ6 range of cars today.

By the late eighties the Toyota Lexus had been launched with a 4 litre V8 engine which set new standards in terms of Powertrain refinement together with very competitive performance. Mercedes Benz introduced their new V8 in 1989, and BMW were known to be developing a 4 litre V8.

It was clear that to remain competitive in the luxury car market that a new range of engines would be necessary by the mid-nineties, and that to achieve class leading objectives necessitated a clean sheet of paper approach.

When we started considering the design concept there was a fairly broad consensus on the configuration, design features and objectives for the engine within the company, but we also felt it crucial that we had an input from potential customers not only to guide the design team but to convince and justify to our management the rationale behind design features and their customer benefits.

To this end we conducted Market Research in all our major markets USA, UK, Germany and Japan and the results were consistent and unambiguous - our customers required:

1. Quality - Class leading reliability and durability were essential.
2. Performance - Not necessarily very high maximum power but excellent performance feel at part throttle and fast response -effortless performance.
3. Refinement - A smooth and quiet engine but with a 'pleasant' sound.
4. Low cost of Ownership - Fuel Consumption and Service costs.

One other interesting outcome was that approximately 20% of respondents, mostly USA, judged engine size by number of cylinders rather than litres or cubic inches. it was also clear that in the USA customers now considered 8 cylinders to be the norm in the premium luxury and sports car segment.

There were other legislative considerations of course which would also have a significant impact on the engine design:

1. Emissions - California continues to adopt increasingly stringent standards within the low Emission Vehicle programme, and in Europe we have the proposals for Stage III.
2. Fuel Consumption. We have CAFE and Gas Guzzler legislation in the USA and proposals in Europe to limit C02 emissions.
3. Noise regulations also continue to become more stringent.
4. Safety - the need to design as compact an engine as possible in order to give freedom to optimise the body structure for crash characteristics.

It was the input from this market research that led us to what became our objective or mission for the programme of 'Refined Power' which was intended to reflect the respondents perception of performance and refinement.

Finally, we had to design and produce the engine within specific investment and design cost constraints to ensure we had a viable business case for the programme.

Having established some broad requirements it was necessary to set objective targets against all the definable engine characteristics, or functional objectives as we call them. This was achieved by benchmarking those vehicles which had been defined as our competitors - Lexus, BMW, Mercedes Benz, Porsche and Infiniti. Defined vehicle attributes such as Performance Feel, Refinement and Driveability were rated on 1-10 scale ranking system for all these cars

Following a review of this data we selected BMW as the benchmark for performance and Lexus for refinement. The engines from these cars were subsequently instrumented and tested on engine dynamometers and in semi-anechoic chambers to establish objective subsets of data against which to compare our own design.

Performance feel for instance was defined by a number of characteristics such as target torque/power curve and rate of change of acceleration against speed.

ENGINE CONCEPT

It is fairly apparent that the first issues that had to be considered were the engine capacity and number of cylinders. We believed that by the use of new technologies to improve engine efficiency and reducing weight that we could best achieve the challenging performance and economy targets by retaining the same engine capacities as in our current range of approx 3.2, 4 and 6 litres.

To reduce the cost base it was proposed to adapt a modular design concept where each engine capacity had the same bore and stroke but different numbers of cylinders. Such an approach gives the benefit of developing only one combustion system, a saving on development time and resource, and common components such as pistons, conn-rods, valve gear and valve train drive components for lower investment and piece cost. Cylinder head and block could also have significant design similarity.

Based on the sales volumes of our current engine range it was anticipated that the highest volume version would be the 4 litre and the choice of 8 cylinders was based on the observations I made earlier, and fits well with a 500cc cylinder volume giving 3 litre V6 and 6 litre V12 engine capacities.

The initial design studies were based on this assumption, but to achieve equal firing intervals and best balance condition on all versions, a 90 degree V8 and 60 degree V6 and V12 included bank angles were required. It soon became apparent, however, that Jaguar's deteriorating financial position was not compatible with such an ambitious engine programme and at this stage in our delibrations the company was purchased by Ford in 1990.

Naturally, one of Ford's first actions was to perform a very exhaustive review of our New Model Cycle Plan. From the engine perspective it was apparent that the demand for V12 engines was declining and that with the global focus on C02 reduction and vehicle emissions in general, our Marketing department's projection for V12 sales could not support the required investment. It was also believed that as we do not sell the current 3.2 litre engine in the USA, and that it represented less than 20% of our total 6 cylinder sales that we should concentrate on the 4 litre V8 and leave the options open for producing a 3 litre V6 or a smaller capacity V8.

Having established what we required we now had the rather more challenging task of achieving it. I'll cover first the engine performance characteristics, then refinement and finally a little on engine management.

PERFORMANCE

With the cylinder capacity determined at 500cc the next major decision required was the bore/stroke ratio.

This graph Fig.1 shows the trade off between bore/stroke ratios and a number of performance factors. Firstly, it can be seen that the larger bore/stroke ratio gives the highest power potential and a shorter stroke gives the potential for lower vibration levels due to the inherently stiffer crankshaft design.

The larger bore configurations, however, have the worst potential for low speed torque and fuel consumption. They also have the higher HC emission potential due to the larger ring circumference available to trap unburned fuel.

Given these conflicting responses we selected a square ratio of 86mm x 86mm as the best compromise guided by our goal of refined power and the need for an environmentally acceptable engine.

From our historical experience it was evident that we required at least 4 valves per cylinders, but a 5 valve design, 3 intake and 2 exhaust was seriously considered and tested. These graphs Fig.2 show the results of those tests. it can be seen that the 5 valve design offered no performance advantage whilst on the right a graph of specific fuel consumption versus EGR percentage, shows a slight advantage for the 4 valve engine at high EGR dilution levels.

It was concluded from this, therefore, that there was no technical justification for the added cost and complexity of 5 valves.

The twin cam 4 valve cylinder head is an evolution of our current 6 cylinder AJ16 design having a very compact combustion chamber with no squish in the interest of HC reduction. The compactness is achieved through the adoption of a narrow included valve angle of 28 degrees and a flat top piston. This chamber design in combination with the controlled metal temperature achieves a high degree of detonation resistance allowing a high compression ratio of 10.75:1 to be used.

This combustion system design gives a high thermal efficiency and high dilution tolerance to EGR for low emissions and fuel consumption.

The high compression ratio in combination with optimised induction port generated swirl results in good combustion noise characteristics by control of burn rate. Engine performance of course is primarily dictated by volumetric efficiency. The volumetric efficiency in turn is significantly affected by valve timing and the optimum varies significantly as a function of speed.

To address this problem we have incorporated what we call a Variable Cam Phasing system which allows a fixed inlet valve period to be optimised in two states Fig.3. This results in not only improved performance but reduced emissions and improved idle quality.

This slide Fig.4 shows volumetric efficiency plotted against engine speed with the inlet cams in the advanced and retarded state. The significant effect on volumetric efficiency is apparent and consequently the timing of this phase change is electronically programmed to take advantage of these characteristics.

In addition the VCP gives a significant reduction in NOx emissions at part throttle conditions. This slide Fig.5 shows the effect on NOx emissions on the left and HC emissions on the right and in each case emissions are plotted against EGR dilution levels. NOx emissions are reduced by the higher exhaust gas residuals with the camshaft in the advanced high overlap condition. This 'internal' EGR reduces NOx formation by reducing maximum combustion temperatures. in addition the HC emissions are also reduced by a small amount due to the re-combustion of the end gasses which would otherwise have passed unburned directly to exhaust. Fig.6 summarises these effects.

The intake and exhaust system designs also have a significant effect on volumetric efficiency. The induction system was therefore designed with long ram pipes tuned for mid-range speeds to compliment the boost from the VCP system at low and high speed conditions.

A plastic inlet manifold was specified for its light weight combined with excellent thermal insulation and smooth internal surface finish. Advantage was also taken of this material to integrate the fuel rail into the manifold, and the bottom feed injectors help prevent fuel vaporisation within the system.

The reduction of friction is a key design requirement to achieve good performance and economy. The lightweight valve gear contributes significantly to this objective, particularly at low engine speeds. In addition the use of Nickasil bore plating and low piston ring loads, high efficiency lubricant and coolant pumps, as well as attention to the bearing design and sizes, has resulted in an engine with very competitive friction levels. This slide Fig.7 shows friction MEP against rpm for our own engines and two 4 litre V8 competitors.

It is the combination of the above features which enabled the performance objective to be achieved Fig.8 of 290 bhp (216 Kw) at 6100 rpm and a peak torque of 290 lbs ft (393 Nm) at 4250 rpm. Over 80% of this peak torque is available from 1400 rpm to 6400 rpm the entire effective speed range. This broad torque range contributes significantly to the performance feel objective.

How this performance compares with other engines can be seen on this slide Fig.9 both in terms of power/litre, peak torque/litre, and torque/litre at 1500 rpm.

The detrimental effect on fuel consumption of cold starts is apparent to us all, but a fast warm up characteristic also reduces emissions and the time for the heater to become effective. To achieve this fast warm up, an innovative cooling system has been designed, described later, and the twin catalysts are mounted directly on the exhaust manifolds to ensure fast 'tight-off'.

The performance of this precision cooling system and close coupled catalyst design are shown on this slide Fig.10 showing the USA emission test cycle. This is a graph of speed against time from a cold start and shows the catalyst to be up to working temperature after 25 secs and the coolant up to its operating temperature after 225 secs.

REFINEMENT

Over the last decade or so there have been a number of designs which attempt to isolate the crankshaft and combustion system from the exterior surfaces of the engine in the interests of reducing noise emissions. An alternative approach is to make the entire engine and transmission assembly as rigid as possible. The former approach was initially investigated but rejected for the following reasons:

1. It is difficult to achieve full isolation - some areas inevitably leak noise.
2. The complex sealing arrangements required were considered to be a risk to quality.
3. Attaching the engine to the transmission and vehicle presented further problems.

We therefore decided to concentrate on designing the most rigid and compact structure possible consistent with light-weight, to reduce the bending deflections, and the lightest possible rotating and reciprocating components.

We believed we could obtain the stiffest structure by using a bedplate with the joint between block and bedplate joint on the crank centre-line, rather than a deep skirt design as used on our V12 and 6 Cylinder engines.

Finite element analysis played a key role in the programme. In the early stages simplified models were used to assess various concept alternatives and help guide our decisions on major features. The first detail component designs were then analysed using FE techniques to ensure robustness prior to physical testing. Based on data from these tests in conjunction with more detailed analysis model were finally used to optimise the performance of the complete powertrain for the key attributes of refinement, durability and sealing performance.

This slide Fig. 11 shows the complete engine FE model and the longitudinal webbing which ties into the torque converter housing.

The precision cooling system, referred to earlier, was developed using CFD and FEA techniques and clearly has a critical role in both cylinder head and block design. The system uses high local coolant velocities and low coolant volume to give both an improvement in control of metal temperatures and fast engine warm up. CFD analysis was used to optimise the flow velocities to ensure good heat transfer throughout the coolant jacket and in conjunction with an FE model predict thermal stress and eliminate thermal fatigue. The system utilises end to end flow of coolant through both block and head. This slide Fig. 12 shows the coolant flows from the water pump at the front of the engine into both banks of the cylinder block where it splits into a by-pass rail and cylinder bore cooling passage. At the rear of the block the flow re-joins and all the coolant returns forward through the cylinder head and thence to the radiator.

The coolant volume of the engine water jacket is only 2.8 litres, being less than 60% of some competitor engines.

The cylinder block is produced from a low pressure die casting as sand cores were considered necessary to optimise the design.

The bedplate is a high pressure aluminium die-casting and incorporates cast in iron bearing inserts to reduce thermal expansion of the crankbores and help maintain tight bearing clearances.

Full powertrain dynamic models including auxiliaries and the transmission unit were used to predict the major vibration modes.

For V8 engines the most significant forcing input for powertrain bending is from the valvegear, whilst for torsion it's from combustion loads. In the AJV8 the valvegear excitation is approx 170 Hz and the combustion loads up to 420 Hz. By optimising the structure to achieve bending frequency over 200 Hz and torsional frequency over 500 Hz we have avoided NVH problem associated with these major modes. Fig. 13 and Fig.14 illustrate these bending and torsional modes.

The features which have contributed to these high natural frequencies are:

1. The bedplate concept.
2. A cast stiffening membrane forming a bridge between the cylinder banks.
3. The light-weight of the cylinder heads.
4. The full diameter torque converter housing which is bolted to the structural sump casting which in turn incorporates extensive cross bracing Fig.15.
5. The compactness and light-weight of the assembly as demonstrated in Fig.16.

A key feature of the cylinder block is the use of Nickel Silicon Carbide electroplated bores. The resulting surface, known as Nikasil, is extremely hard, giving a very durable bore condition. A smooth surface finish less than 3 Ra allows the use of low radial load ring pressures. Bore distortion is reduced by the elimination of the differential expansion characteristics of an iron liner in an aluminium block.

The lightweight piston design has a top land depth of only 4.9 mm and 29.9 mm compression height which helps achieve a weight of under 450 gms. It is a monometal construction and the top compression ring is molyedenum plated.

The con-rod is sinter forged and the process accuracy is such that no balance correction masses are required. The con-rod cap is fracture split, after machining, which besides reducing machining gives very accurate and secure re-location.

The crankshaft was designed to compliment the cylinder block structure and spheroidal graphite cast iron was used to give the best combination of: freedom of design; weight; machineability; internal damping; and cost.

The bearing diameters were chosen to minimise friction consistent with adequate stiffness. The effective bearing surface area was maximised by using unusually small fillet radii. Variable pressure fillet rolling was used to increase the pressure in the pin overlap area to give over 50% improvement in strength compared to an unrolled part. The balance weights are arranged to minimise bearing loads and maintain crankshaft dynamic balance at high engine speeds.

The benefit of weight reduction on valve gear components was considered crucial in so far as it affects total engine weight, engine friction and refinement. To this end the valve stems diameter are only 5 mm, and the bucket tappets are aluminium, where historically we have used iron or steel. Direct acting cams with mechanical adjustment was adopted instead of hydraulic tappets favoured by many of our competitors in the interests of weight reduction and reliability Fig.17. In order to achieve our performance objective the inlet valves are larger in relation to the bore size than our competitors to improve performance potential.

Fig.18 shows the resulting valve gear weight compared with other engines of similar size, and also valve diameter as a function bore size.

ENGINE MANAGEMENT

Engine Management Systems continue to become more complex largely in response to increasingly stringent emissions control standards and diagnostic requirements although significant improvements are also being made in reliability, engine performance and economy.

The chosen system supplier for the AJ-V8 was Denso who were given total responsibility for the system design with Jaguar responsible for calibration and system validation. The Electronic Control Module and several major components specifications were developed exclusively to meet Jaguar's requirements, including the requirement to meet worldwide emission standards.

The system was developed over a four year period with work being undertaken at Denso headquarters in Japan, at their purpose built laboratory in Coventry, and our Engineering centre at Whitley.

Hot and Cold climate work was performed in Arizona and Ontario and high altitude in Colorado.

Further high speed testing was undertaken at Nardo (Italy) and Pecos (Texas).

The fuel management part of the system includes the main input and output components listed on this slide Fig.19, and the many software strategies ensure that fuelling is closely controlled under all driving conditions.

The twin spray injectors are located close to the inlet valves in order to optimise starting and transient response by reducing inlet wall 'wetting'

Air flow is sensed by a hot wire mass air flowmeter and coolant temperature and barometric pressure sensors modify the air/fuel ratio as a function of engine temp during warm up and ensures it remains constant under varying atmospheric conditions.

Fuelling disablement plays a major role in both torque reduction for Traction and Stability Control and the need for component protection in the event of system malfunction.

A major function of the Fuel Injection system is the control of Air/Fuel ratio within the very fine limits required to achieve Californian emission standards in particular.

In order to achieve this a total of four oxygen sensors are used, one situated upstream and one downstream of the 2 litre close coupled catalyst on each cylinder bank.

The upstream sensor is the primary feedback sensor, controlling the fuelling during normal driving conditions while the downstream sensor provides a longer term adaption of the basic calibration over the vehicle life. It is also used for monitoring the efficiency of the catalyst for diagnostic purposes.

This feedback control system is completely independent on each bank of four cylinders.

This next slide Fig.20 shows why we need such accurate control of air fuel ratio for emission purposes. This simplified diagram shows CO and HC can readily be oxidised to C02 and water when air fuel ratio is lean and how NOx can easily be reduced when the mixture is rich. The air fuel ratio must be controlled in the very narrow regime where both oxidisation and reduction occur simultaneously. An accuracy of better than (0.5% is necessary to prevent significant loss of conversion efficiency.

Turning to the Ignition part of the system. Each cylinder is provided with an individual coil mounted over the spark plug thus eliminating the need for HT leads. They also have the benefit of allowing large swings in ignition timing enabling this to be used as a very effective and highly responsive means of torque control.

A further dynamic control strategy is combustion knock control. Dual cylinder block mounted accelerometers transmit data to the ECU where complex filtering and gating, together with statistical data processing algorithms enable ignition timing to be continuously adjusted on each cylinder individually to avoid knock.

This feature optimises the ignition timing for torque and economy at all conditions over the life of the engine and also caters for different quality fuels.

An electronic throttle was part of the original EMS specification and includes integrating all the functions listed on this slide Fig.21.

The first five of these functions are features of most current cars, but significant benefits are apparent from their integration. Whilst some, idle speed and additional air are controlled by the EMS others such as cruise control and traction control frequently have their own electronic control unit and supplier. Hardwired connections are necessary and co-ordinating the many control strategies is not an easy task.

By combining all the functions within one actuator and having the software integrated in one ECM and written by one supplier considerably reduced these problems and improves quality.

Dynamic throttle position filtering and fine tuning of throttle pedal progression are new features enabled only by an electronic system.

In addition the throttle body is shipped to our engine plant as a complete assembly. No adjustments are required for any of the functions over the engines life and assembly time is reduced.

The system also controls the auxiliary functions listed on this slide Fig.20. The control of such items as fans, air con, clutch, etc allows these to be enabled or inhibited dynamically to assist driveability.

The EMS communicates with other vehicle systems to enhance vehicle performance. Engine torque can be adjusted, up to 30 times per second under direct control of the Transmission during gearshifts in order to improve shift quality and improve durability.

The traction control system can also request torque reduction from the engine and depending on the prevailing conditions this can be achieved by ignition retard, fuelling or throttle intervention.

I won't go into any detail on the emission control system other than to say it is designed to comply with the most stringent California TLEV standards. To achieve this the catalyst is operating and feedback control initiated after only 25 secs from the test start. These features together with the close control of air fuel ratio give a hydrocarbon conversion of up to 98% over the emission test including the cold start.

In summary, therefore, the EMS plays a primary role in vehicle functionality with respect to Emissions Economy, Driveability, Performance, Safety and Reliability.

The AJV8 as now in production for our XK8 has fulfilled its design objective and demonstrates a number of Best in Class features for:

Specific Performance	72.5 HP/Litre
Weight	200 Kg
Stiffness	>200 Hz
Cooling System	4 minute warm up
Service Costs	10,000 miles
	30,000 miles

It only remains for me to acknowledge the dedication, skill and teamwork demonstrated by all the personnel involved in the engine which I have described this evening.

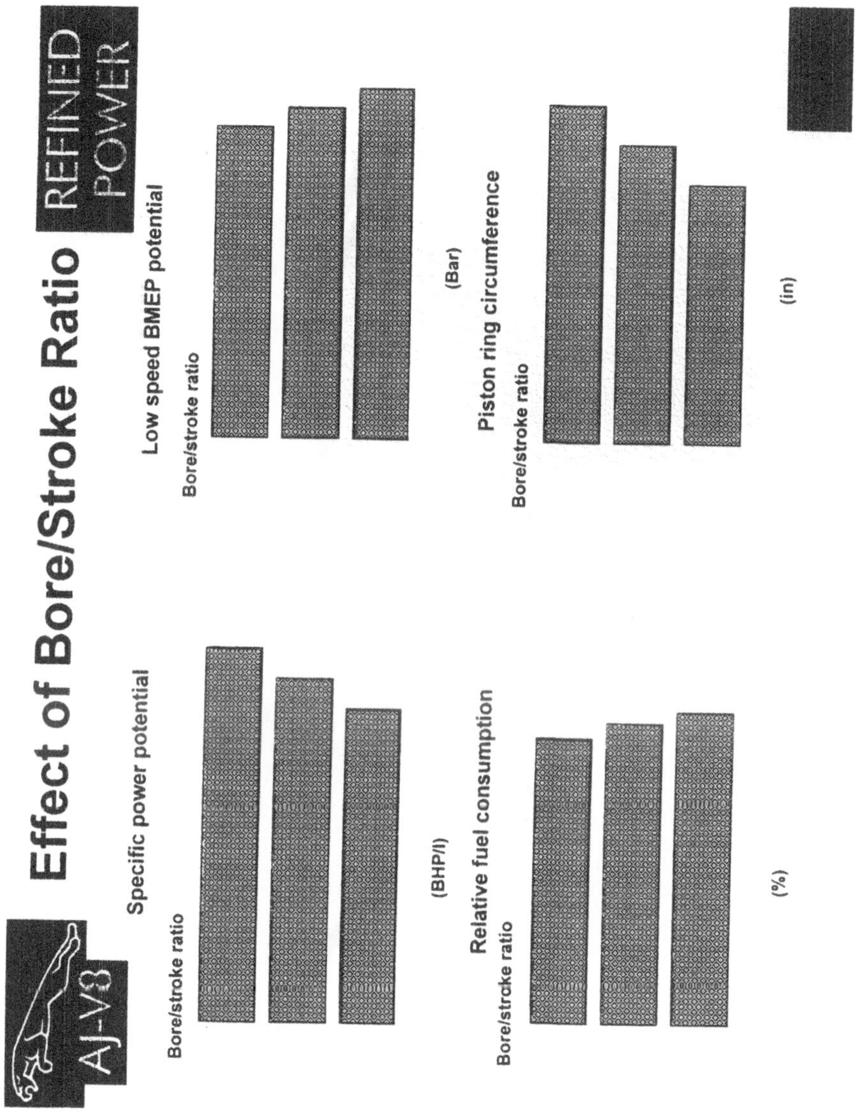
AJ-V8
Effect of Bore/Stroke Ratio
REFINED POWER
Specific power potential
Bore/stroke ratio
(BHP/l)
Low speed BMEP potential
Bore/stroke ratio
(Bar)
Relative fuel consumption
Bore/stroke ratio
(%)
Piston ring circumference
Bore/stroke ratio
(in)

Fig 1

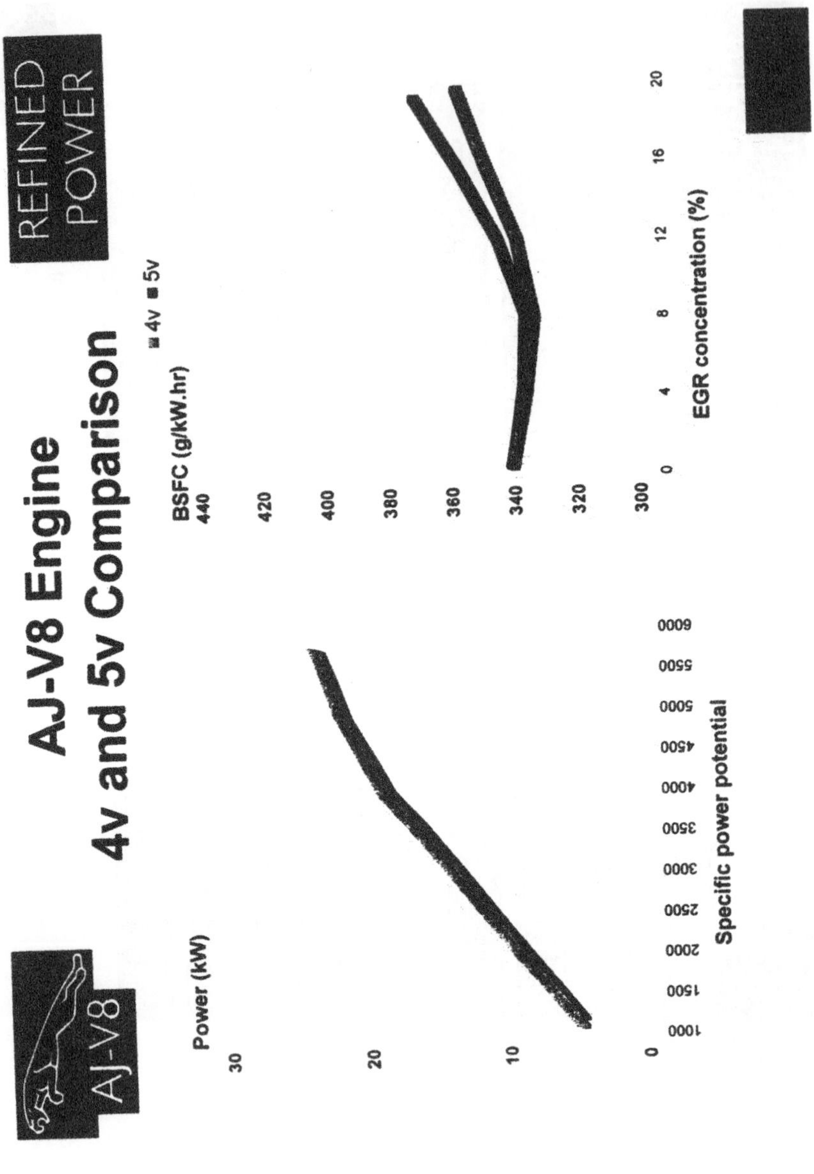
AJ-V8 Engine
4v and 5v Comparison
AJ-V8
REFINED POWER
Power (kW)
30
20
10
0
1000
1500
2000
2500
3000
3500
4000
4500
5000
5500
6000
Specific power potential
BSFC (g/kW.hr)
440
420
400
380
360
340
320
300
0
4
8
12
16
20
EGR concentration (%)
4v
5v

Fig 2

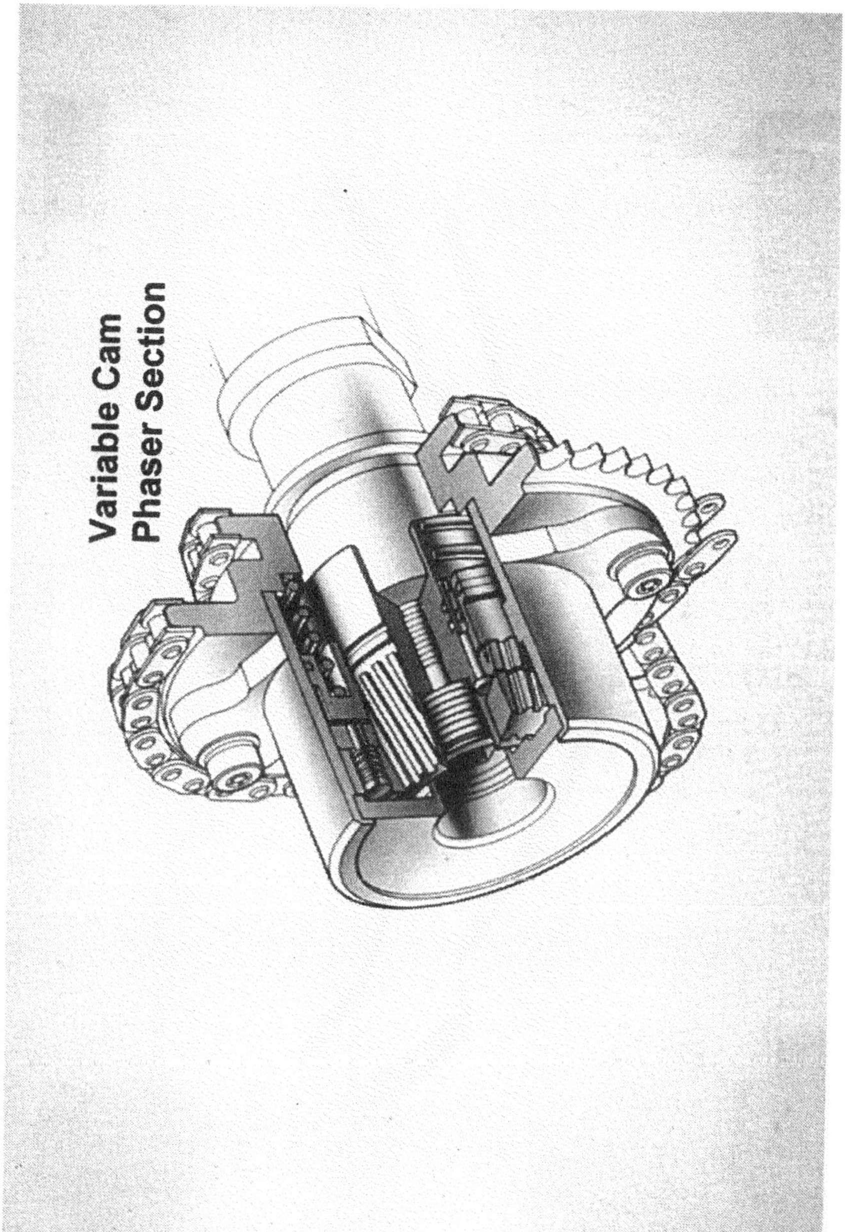

Fig 3

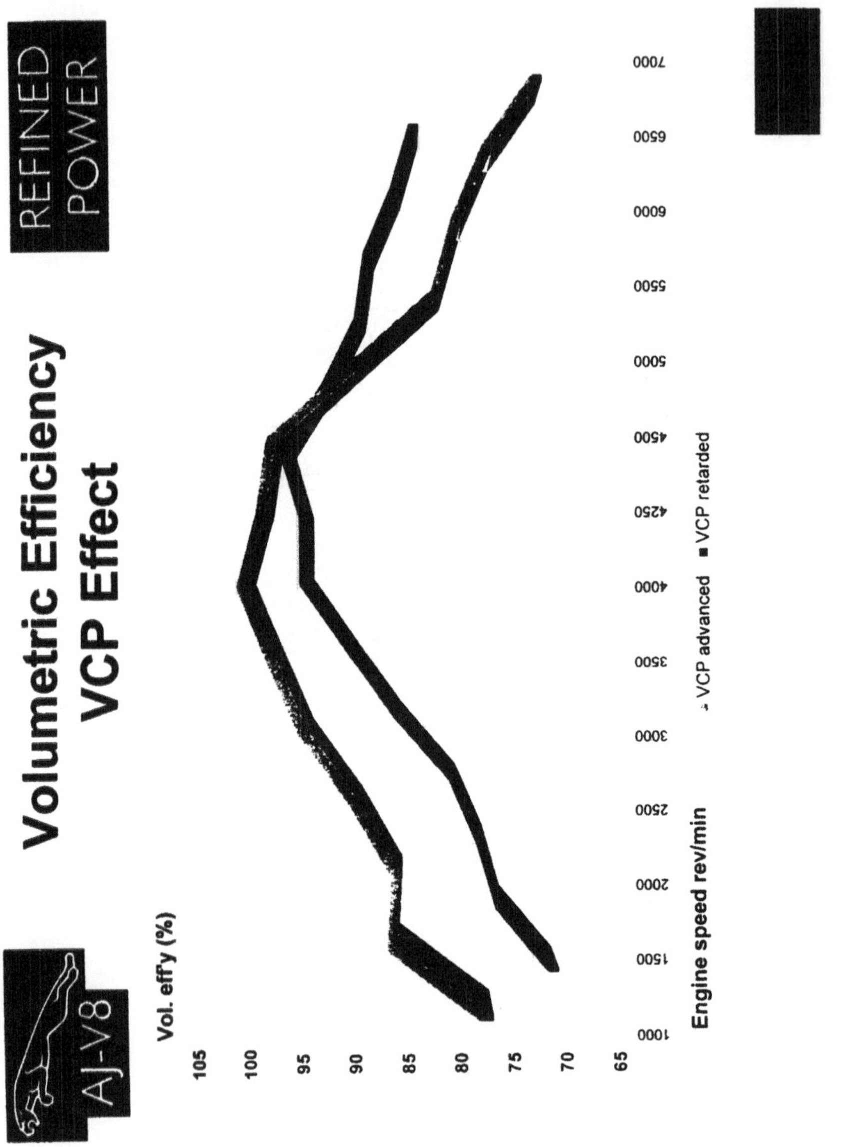

Volumetric Efficiency
VCP Effect
REFINED POWER
AJ-V8
Vol. eff'y (%)
105
100
95
90
85
80
75
70
65
1000
1500
2000
2500
3000
3500
4000
4250
4500
5000
5500
6000
6500
7000
Engine speed rev/min
VCP advanced
VCP retarded

Fig 4

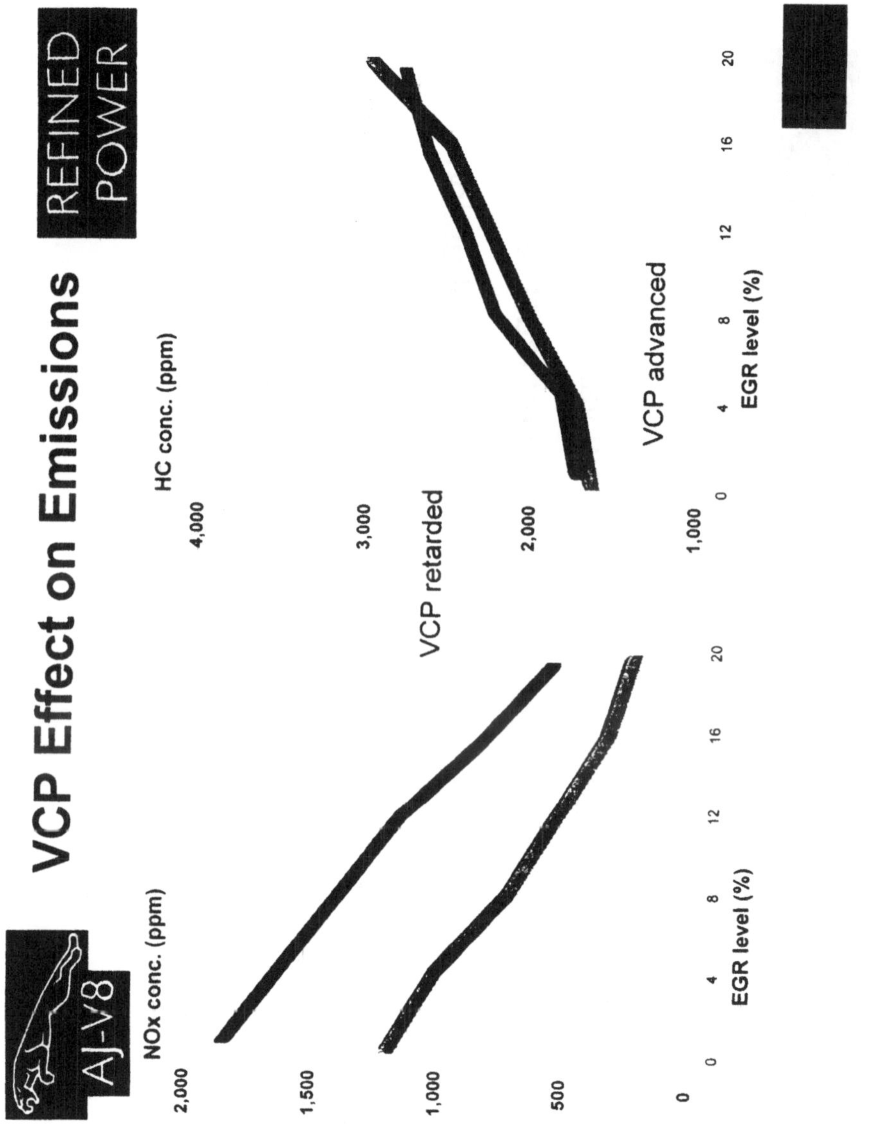

AJ-V8
VCP Effect on Emissions
REFINED POWER
NOx conc. (ppm)
2,000
1,500
1,000
500
0
0
4
8
12
16
20
EGR level (%)
HC conc. (ppm)
4,000
3,000
2,000
1,000
0
4
8
12
16
20
EGR level (%)
VCP retarded
VCP advanced

Fig 5

AJ-V8

VCP Fundamentals

REFINED POWER

- VCP advanced
  - High valve overlap
  - High exhaust residuals
- Reduced emissions
  - NOx and HC
- Improved volumetric efficiency
  - Low speed

- VCP retarded
  - Low valve overlap
  - Low exhaust residuals
- Improved combustion stability
- Improved volumetric efficiency
  - High speed

Fig 6

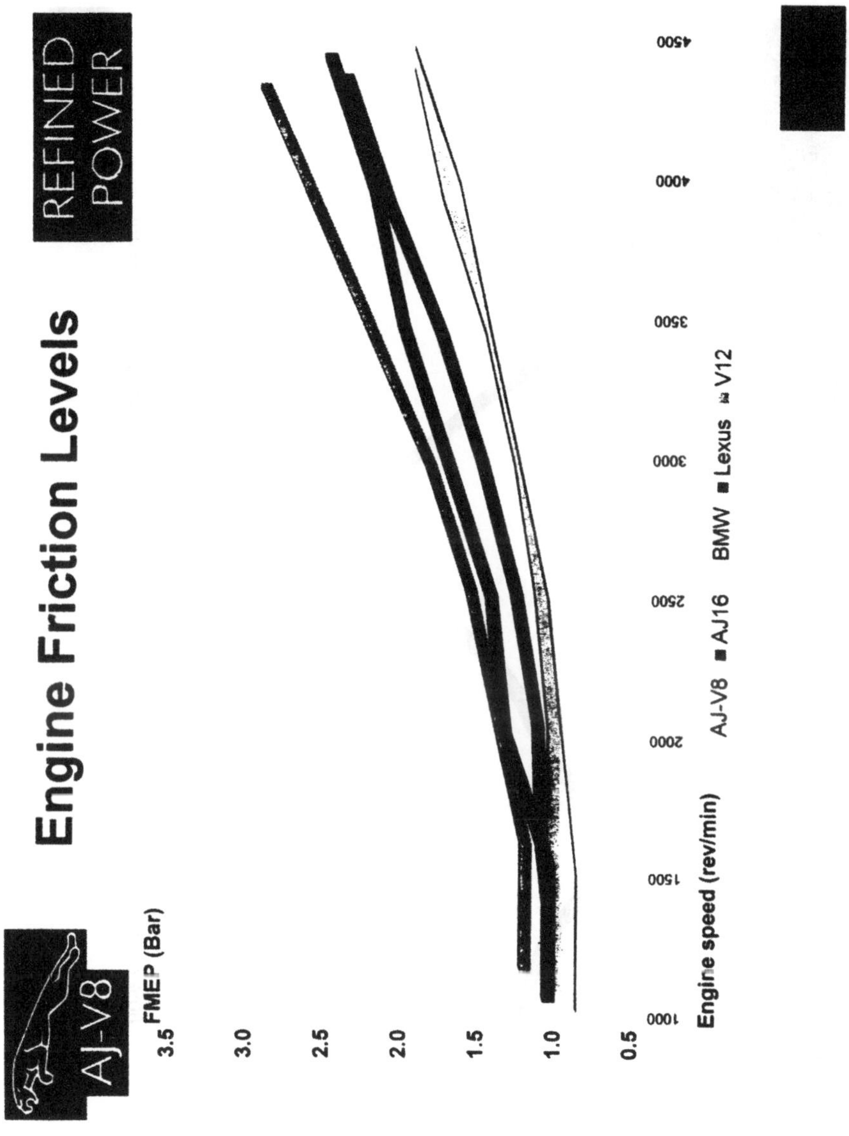
Engine Friction Levels
AJ-V8
REFINED POWER
FMEP (Bar)
3.5
3.0
2.5
2.0
1.5
1.0
0.5
1000
1500
2000
2500
3000
3500
4000
4500
Engine speed (rev/min)
AJ-V8
AJ16
BMW
Lexus
V12

Fig 7

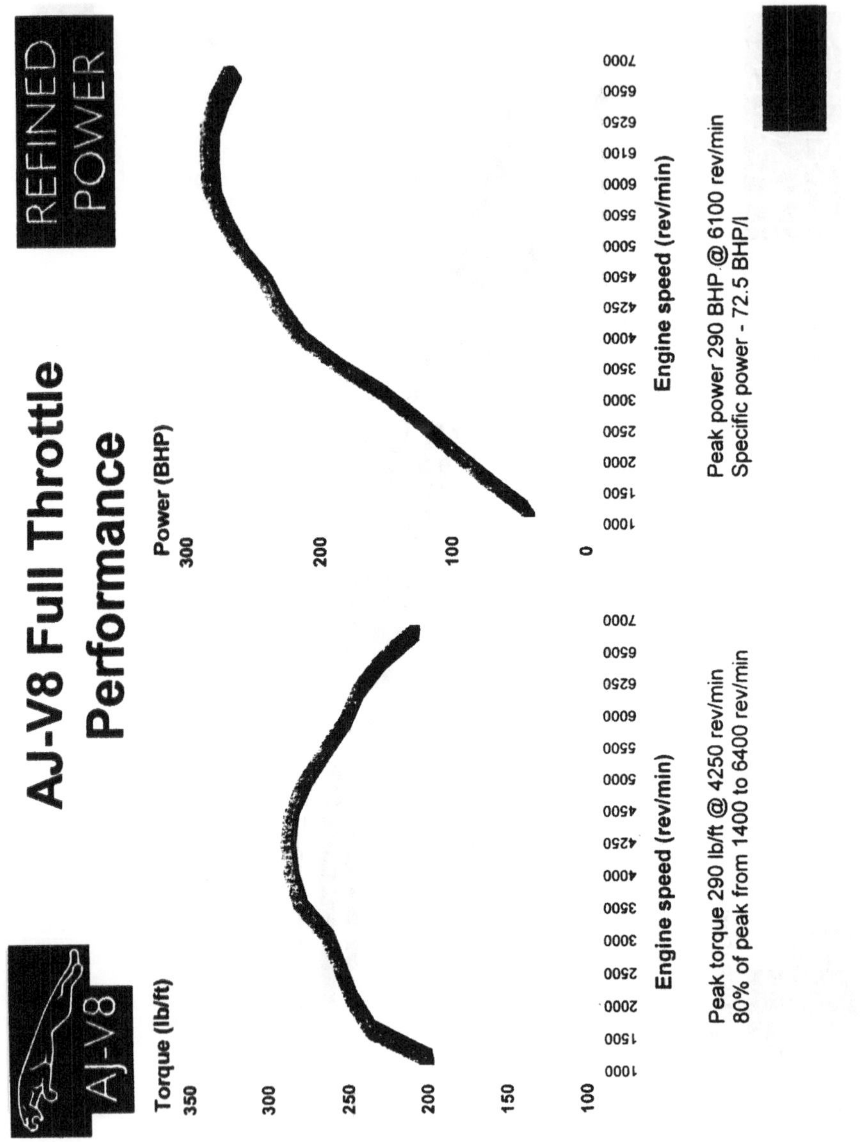
AJ-V8
AJ-V8 Full Throttle Performance
REFINED POWER
Torque (lb/ft)
350
300
250
200
150
100
1000
1500
2000
2500
3000
3500
4000
4250
4500
5000
5500
6000
6250
6500
7000
Engine speed (rev/min)
Peak torque 290 lb/ft @ 4250 rev/min
80% of peak from 1400 to 6400 rev/min
Power (BHP)
300
200
100
0
1000
1500
2000
2500
3000
3500
4000
4250
4500
5000
5500
6000
6100
6250
6500
7000
Engine speed (rev/min)
Peak power 290 BHP @ 6100 rev/min
Specific power - 72.5 BHP/l

Fig 8

Fig 9

AJ-V8

Competitor Performance

REFINED POWER

	Specific power (BHP/l)	Power density (BHP/lb)	Peak BMEP (Bar)	BMEP @ 1500 rev/min (Bar)
AJ-V8	72.5	0.66	12.4	10.0
AJ16	59.4	0.44	12.0	9.9
V12	50.1	0.40	10.1	8.8
Lexus	65.0	0.59	11.5	9.5
BMW	63.9	0.58	12.0	9.5
Mercedes	67.0	0.58	12.3	9.6

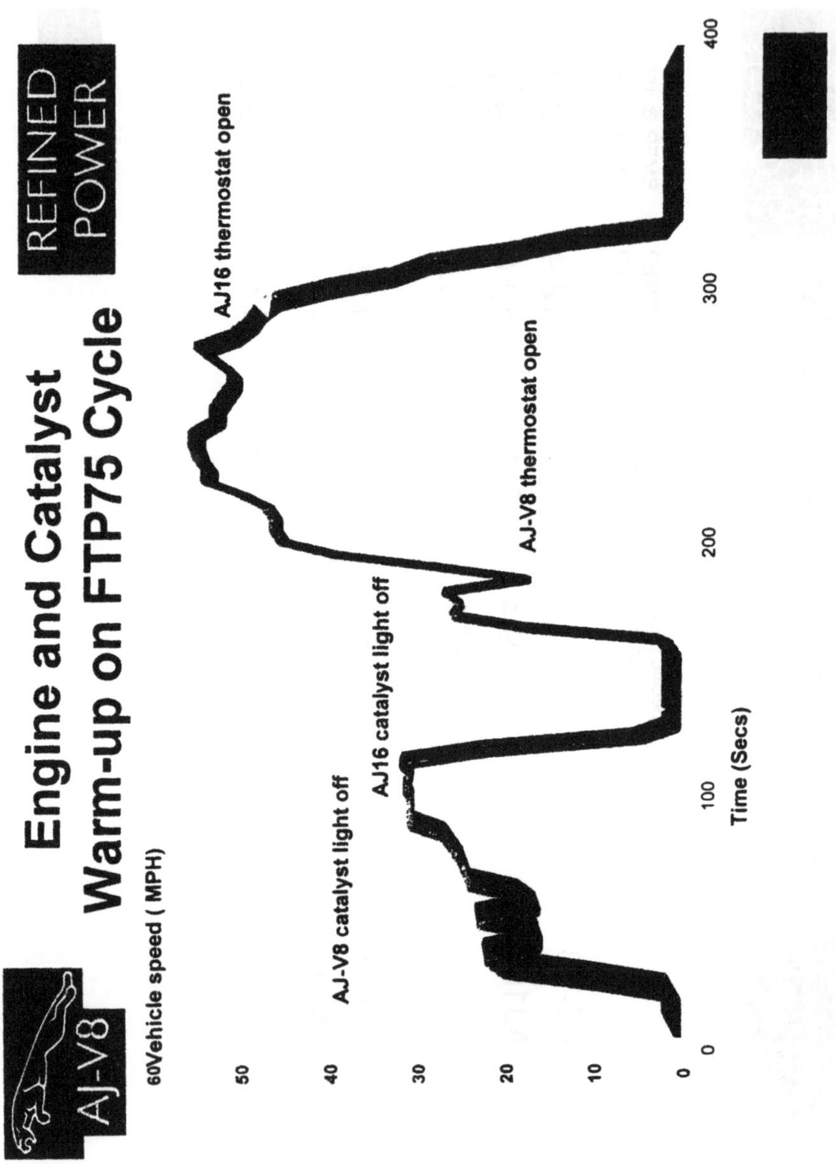
AJ-V8
REFINED POWER
Engine and Catalyst Warm-up on FTP75 Cycle
60Vehicle speed (MPH)
50
40
30
20
10
0
0
100
200
300
400
Time (Secs)
AJ-V8 catalyst light off
AJ16 catalyst light off
AJ-V8 thermostat open
AJ16 thermostat open

Fig 10

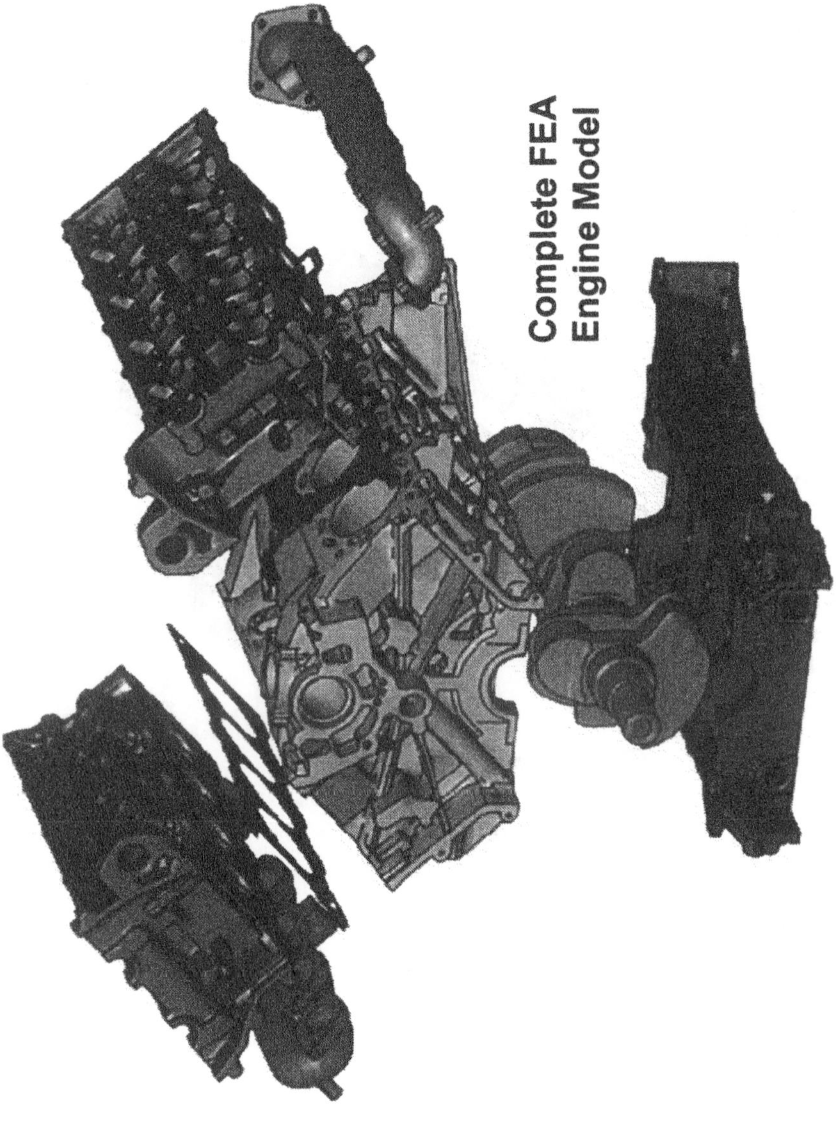

Fig 11

AJ-V8

Features

REFINED POWER

- Precision cooling system

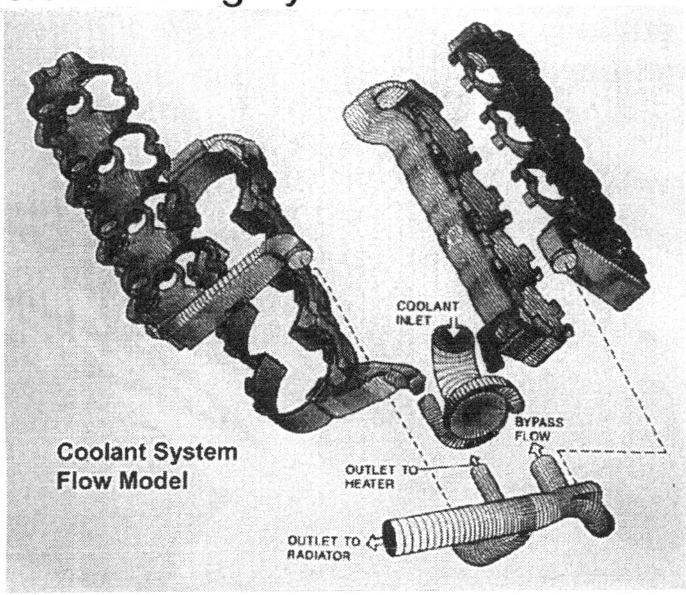

Fig 12

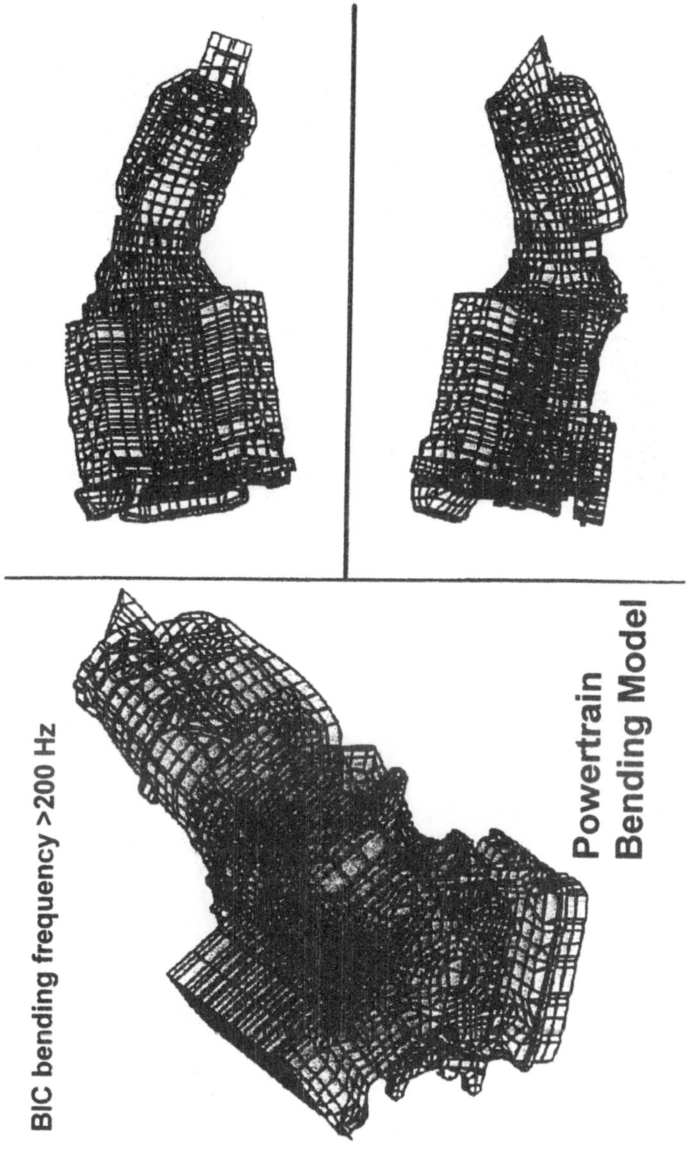

Fig 13

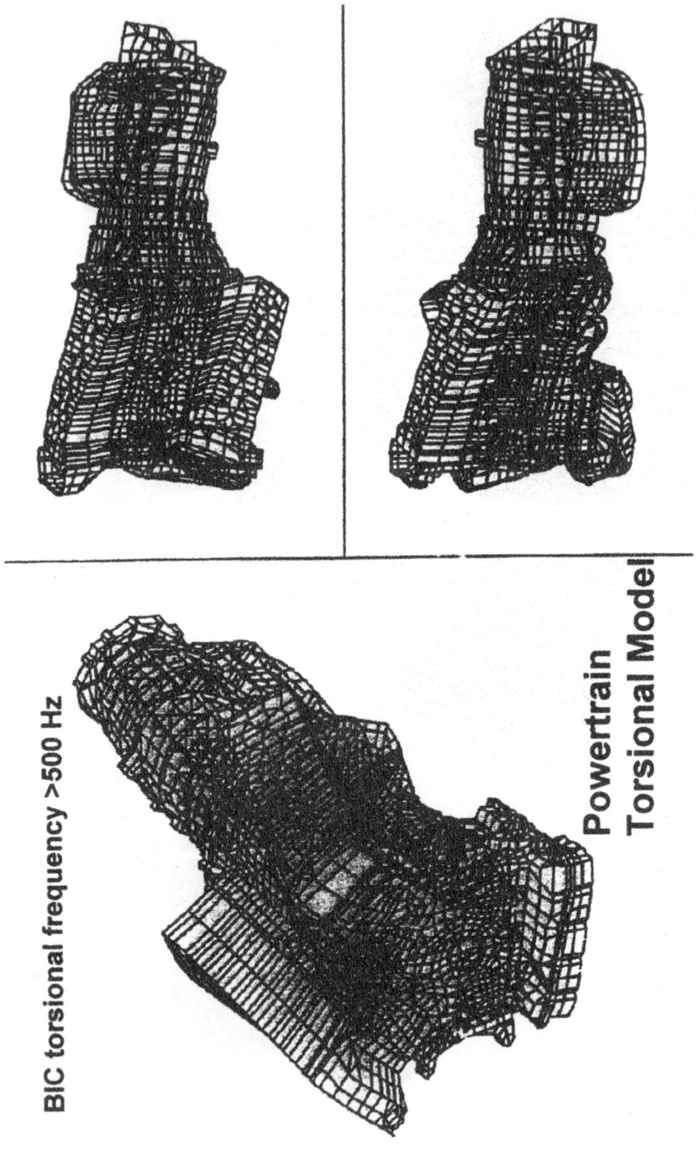
BIC torsional frequency >500 Hz
Powertrain
Torsional Model

Fig 14

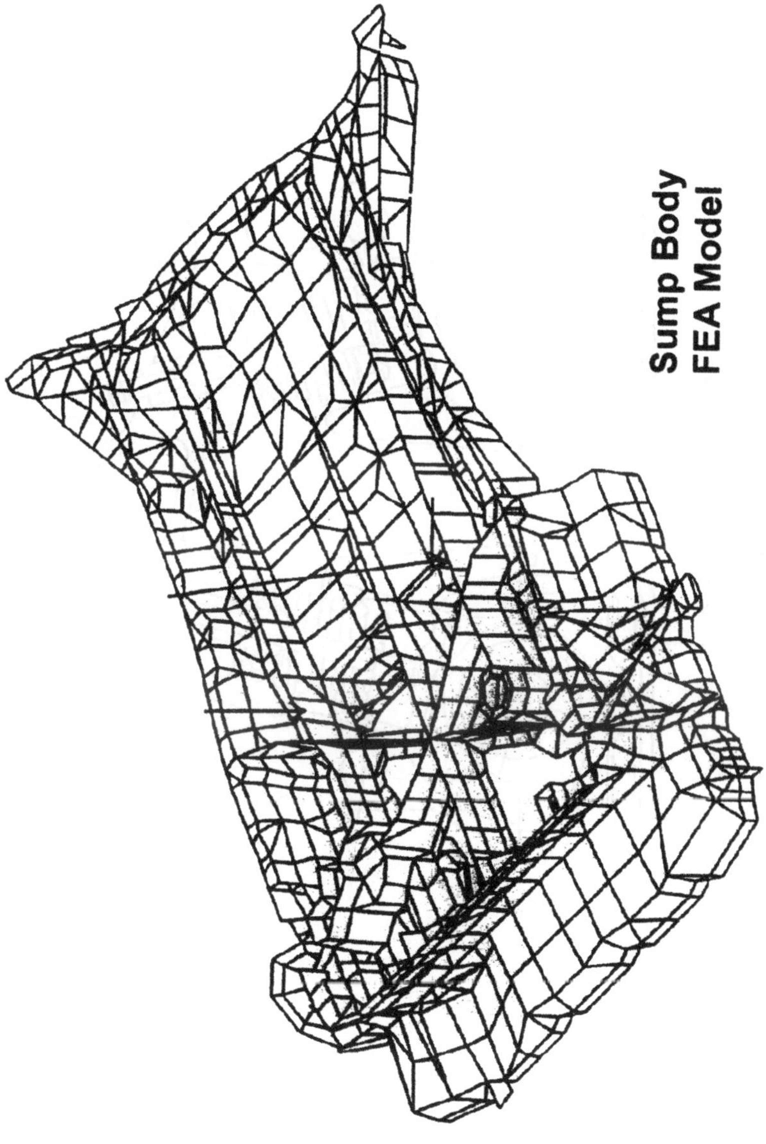

Fig 15

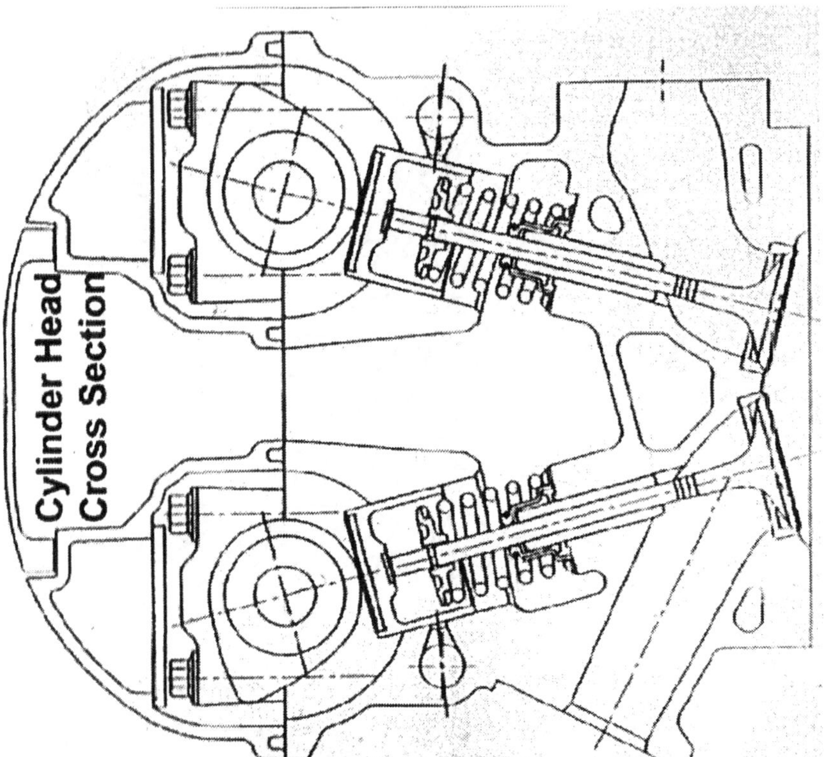

Fig 17

Fig 18

AJ-V8

Valve Gear Comparison

REFINED POWER

	Valve mass inlet/exhaust [g]	Inlet valve ⌀ / Bore ⌀
AJ-V8	96 /96	0.407
AJ16	129 / 130	0.394
V12	163 / 150	0.455
Lexus	109 / 100	0.391
BMW	149 / 147	0.393
Mercedes	146 / 141	0.391

Valve Gear Comparison

AJ-V8 — REFINED POWER

	Valve mass inlet/exhaust [g]	Inlet valve ⌀ / Bore ⌀
AJ-V8	96 /96	0.407
AJ16	129 / 130	0.394
V12	163 / 150	0.455
Lexus	109 / 100	0.391
BMW	149 / 147	0.393
Mercedes	146 / 141	0.391

Fig 18

Fig 19

AJ-V8

Fuel Injection System

REFINED POWER

- Twin-spray phased port injection
- Air flow sensed by hot-wire mass flowmeter
- Coolant temperature input
- Barometric pressure compensation
- Disablement strategies for
  - Torque reduction for traction control
  - Catalyst protection

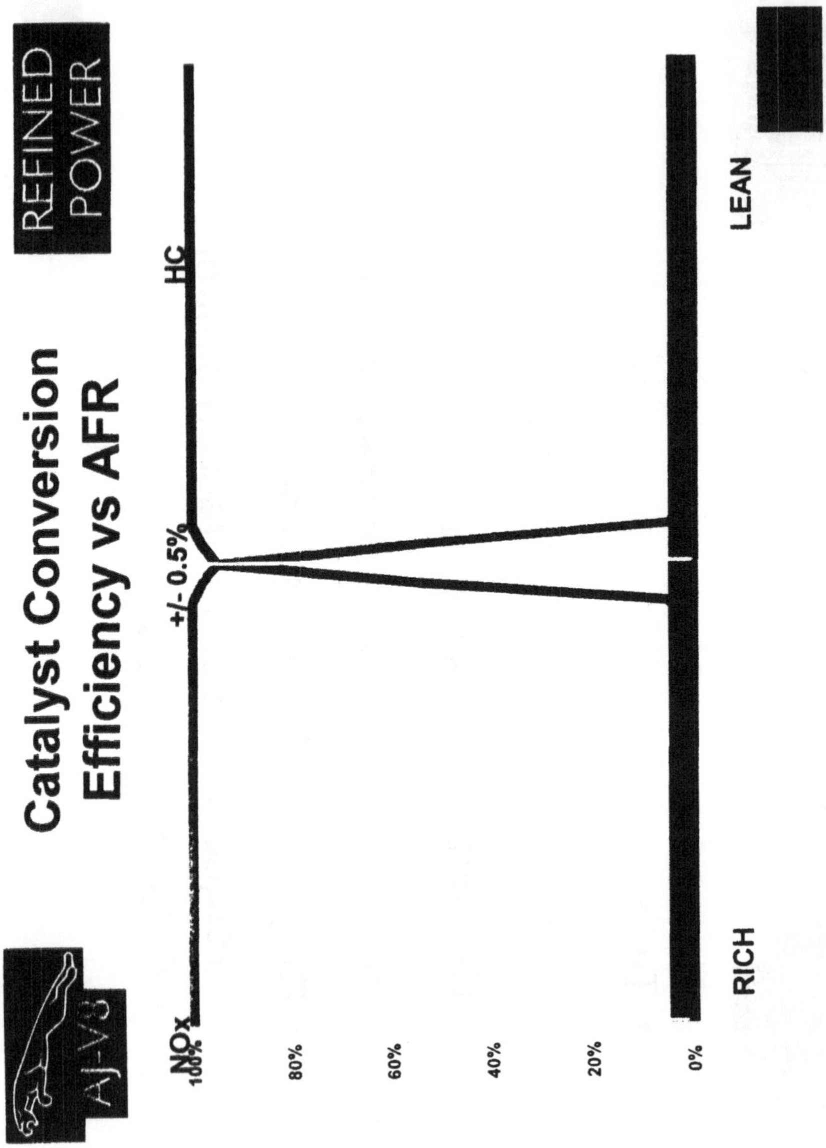
AJ-V8
Catalyst Conversion
Efficiency vs AFR
REFINED
POWER
NOx
HC
+/- 0.5%
100%
80%
60%
40%
20%
0%
RICH
LEAN

Fig 20

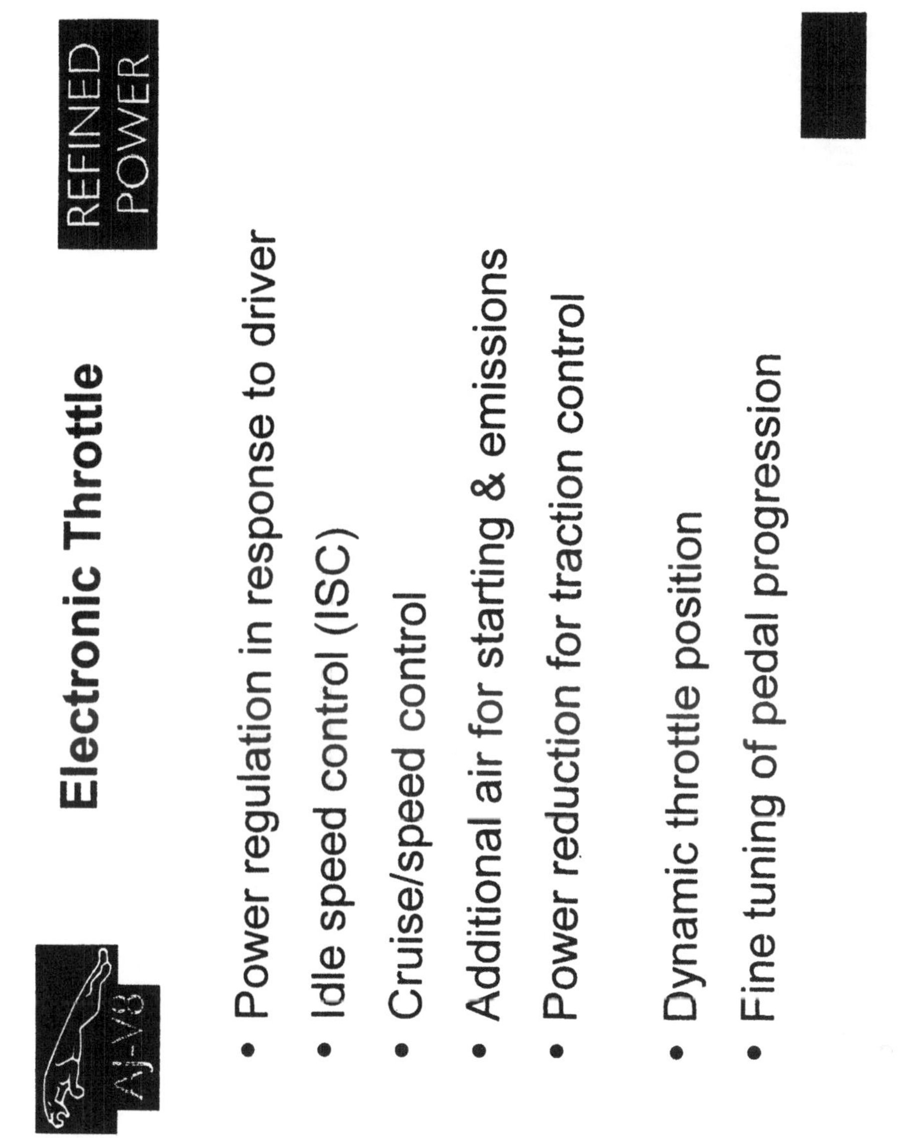
Electronic Throttle
AJ-V8
REFINED POWER
• Power regulation in response to driver
• Idle speed control (ISC)
• Cruise/speed control
• Additional air for starting & emissions
• Power reduction for traction control
• Dynamic throttle position
• Fine tuning of pedal progression

Fig 21

Auxiliary EMS Functions

REFINED POWER

- 2-position mapped VCP
- Electronic EGR valve
- Purge control
- Multi-speed radiator cooling fan
- Air-conditioning clutch control
- Interfacing with vehicle systems via CAN
  - Transmission
  - Traction
  - Instrument pack

Fig 22

Index

Printed and bound by CPI Group (UK) Ltd, Croydon, CR0 4YY
13/09/2026
14961034-0001